ADAMS'S MENSURATION.

# MENSURATION,

# MECHANICAL POWERS,

AND

# MACHINERY.

THE PRINCIPLES OF MENSURATION ANALYTICALLY EXPLAINED, AND PRACTICALLY APPLIED TO THE MEASUREMENT OF LINES, SUPERFICIES, AND SOLIDS; ALSO, A PHILOSOPHICAL EXPLANATION OF THE SIMPLE MECHANICAL POWERS, AND THEIR APPLICATION TO MACHINERY.

DESIGNED FOR THE USE OF SCHOOLS AND ACADEMIES.

NEW YORK:
ROBERT B. COLLINS.
1850.

Stereotyped by
HOBART & ROBBINS;
NEW ENGLAND TYPE AND STEREOTYPE FOUNDERY,
BOSTON.

# PREFACE.

MORE than nineteen twentieths of the children in our country receive all their education in the common schools. And but about one half of the number who attend the high schools and academies, ever go further in a course of mathematical study than through the elements of Algebra and Geometry. Hence, of the whole number of scholars in all the schools in our country, not more than one fortieth ever acquire as much knowledge of the principles of Mensuration as is actually needed for the every-day business of life.

Now, this is manifestly wrong. Every person, and more especially every young man, should possess sufficient knowledge of the principles of Mensuration to enable him to transact his business independent of arbitrary rules, or of the assistance of those who, having been more fortunate than himself in acquiring a knowledge of these principles, render the necessary aid only for a stipulated sum.

But wherein lies this evil? It is not in the want of treatises upon Mensuration; for the world is well supplied, upon this subject, with text-books abounding in mechanical rules. Neither is it in the want of facilities for acquiring a thorough mathematical knowledge; for the doors of our high schools, seminaries and colleges, are open alike to all who may choose to enter. The rules and principles presented in most of the text-books heretofore written upon this department of education, are mere directions for the performance of a mechanical process, which, if followed, will "bring the answer."

The reason generally given for thus presenting them is, that the principles involved cannot be understood without a thorough knowledge of Geometry. But this is not so. A great majority, nay, nearly all of the rules and principles involved in Mensuration as applied to the actual business of life, admit of an analysis perfectly comprehensible by the mere arithmetician.

The evil must be, then, in the want of *the proper kind of text-book;* — one that shall give the *why*, as well as the *how;* one that shall be adapted to the capacity of the student who has no knowledge of mathematics beyond Arithmetic.

Such a work, it is believed, is here presented to the public. The characteristics of the work are the following:

1. *It is an analytical work.* No rule or principle is introduced as mechanical, that admits of an analysis intelligible to the mind of the thorough arithmetical scholar. After the number of rules that admit of such an analysis are taken from the whole number contained in the work, it leaves but *a very small number* of arbitrary mechanical rules.

2. *The arrangement is natural and philosophical.* The subject of Weights and Measures is first considered, for the reason that nothing can be measured without reference to some established standard of weight or measure. The Geometrical Definitions, necessary to be understood by the pupil in pursuing this study, are next introduced; and these are followed by a number of important Geometrical Problems. The Mensuration of Lines and Superficies is then presented, and next in order follows the Mensuration of Solids; care being taken in all cases to present the various rules in their most natural order. The Simple Mechanical Powers are next considered; and the work closes with an application of the Mechanical Powers to machines, and an examination of some of the important principles of Machinery.

4. *The "Topic Method" of questioning*, which was followed in "Adams's Book-keeping," having been received with much favor, is adopted in this work. This method points out something for the pupil to do, and *it also requires him to do it.*

4. In the analysis of the various principles, and in the examples for practice, care has been taken to avoid the extremes of analysis and synthesis. The work is therefore neither so obscure as to be unintelligible to the majority of pupils, nor so puerile as to leave nothing upon which the active and inquiring mind may exercise and improve itself.

5. The analysis of many of the rules and principles, and the peculiar manner in which the subjects generally are presented, are believed to be original. The Encyclopædia Britannica, North American Review, the works of Dr. Lardner, Galloway, Coulomb, Rennie, Willis, and Gregory, and many of the first teachers, machinists and mechanics, in New England and New York, have been consulted in the preparation of the work.

The work contains just the kind of information required by the mass of people throughout the country; and it is confidently hoped that its arrangement, and its adaptation to the best and most approved methods of teaching, together with the importance of the subject, may secure for the work a place in the course of instruction in all our schools and academies, though it be, in some cases, at the expense of some of the higher but less important branches.

# INDEX.

## WEIGHTS AND MEASURES.

### Weights.


I. Troy Weight, . . . . . . . . 9
II. Apothecaries' Weight, . . . . . . . . 10
III. Avoirdupois or Commercial Weight, . . . . . . . . 10


### Measures of Extension.


I. 1. Linear Measure, . . . . . . . . 11
2. Cloth Measure, . . . . . . . . 12
3. Linear Chain Measure, . . . . . . . . 12
4. Duodecimal Linear Measure, . . . . . . . . 12
5. Miscellaneous Linear Measures, . . . . . . . . 12
II. 1. Square Measure, . . . . . . . . 13
2. Square Chain Measure, . . . . . . . . 14
3. Artificers' Superficial Measure, . . . . . . . . 14
III. Cubic or Solid Measure, . . . . . . . . 14


### Measures of Capacity.


I. Wine Measure, . . . . . . . . 15
II. Beer Measure, . . . . . . . . 16
III. Dry Measure, . . . . . . . . 17
Standard Road Measures, . . . . . . . . 18


## DEFINITIONS.

### Geometrical Definitions.


Lines and Angles, . . . . . . . . 21
Plane Figures, . . . . . . . . 23
Rectilinear Plane Figures, . . . . . . . . 23
Curvilinear and Mixtilinear Plane Figures, . . . . . . . . 27
Solids or Bodies, . . . . . . . . 30


## PRACTICAL GEOMETRY.

### Geometrical Problems.


Prob. 1. To draw a line parallel to a given line, . . . . . . . . 35
Prob. 2. To bisect a given line, . . . . . . . . 36
Prob. 3. To bisect a given curve, . . . . . . . . 36
Prob. 4. To bisect a given angle, . . . . . . . . 36
Prob. 5. To erect a perpendicular on the middle of a given line, . . . . . . . . 37
Prob. 6. To erect a perpendicular on any given point in a line, . . . . . . . . 37
Prob. 7. From any point without a given line to draw a perpendicular to the line, . 37
Prob. 8. To describe a circle which shall pass through any three given points not in a right line, . . . . . . . . 38
Prob. 9. To find the center of a circle, . . . . . . . . 38
Prob. 10. To find the center of a circle of which an arc only is given, . . . . . . . . 38
Prob. 11. To draw a curve parallel to a given curve, . . . . . . . . 39
Prob. 12. To describe a right-angled triangle, the base and perpendicular being given, . 39
Prob. 13. To describe an equilateral triangle, . . . . . . . . 39

Prob. 14. To describe a triangle, the three sides being given, . . . . . . . . . . . . 39
Prob. 15. To describe a right-angled triangle, the hypotenuse and one side being given, . 40
Prob. 16. To make an angle equal to a given angle, . . . . . . . . . . . . . . . . 40
Prob. 17. To describe a triangle, two sides and the angle which they contain being given, . . . . . . . . . . . . . . . . . . . . . . . . . . . . . 40
Prob. 18. To describe a square, . . . . . . . . . . . . . . . . . . . . . . . . 40
Prob. 19. To describe a rectangle, . . . . . . . . . . . . . . . . . . . . . . 41
Prob. 20. To describe a rhombus, . . . . . . . . . . . . . . . . . . . . . . 41
Prob. 21. To inscribe a triangle in a circle. . . . . . . . . . . . . . . . . . . 41
Prob. 22. To inscribe a square in a circle, . . . . . . . . . . . . . . . . . . 42
Prob. 23. To inscribe a pentagon in a circle, . . . . . . . . . . . . . . . . . 42
Prob. 24. To inscribe a hexagon in a circle, . . . . . . . . . . . . . . . . . . 42
Prob. 25. To inscribe an octagon in a circle, . . . . . . . . . . . . . . . . . . 42
Prob. 26. To inscribe a decagon in a circle, . . . . . . . . . . . . . . . . . . 43
Prob. 27. To inscribe a dodecagon in a circle, . . . . . . . . . . . . . . . . . 43
Prob. 28. To inscribe any regular polygon in a circle, . . . . . . . . . . . . . . 43
Prob. 29. To describe any regular polygon, . . . . . . . . . . . . . . . . . . 43
Prob. 30. To circumscribe a regular polygon about a circle, . . . . . . . . . . . . 44
Prob. 31. To circumscribe a circle about a regular polygon, . . . . . . . . . . . . 44
Prob. 32. To inscribe a circle in a regular polygon, . . . . . . . . . . . . . . . 44
Prob. 33. To inscribe a circle in a triangle, . . . . . . . . . . . . . . . . . . 44
Prob. 34. To construct solids, . . . . . . . . . . . . . . . . . . . . . . . . 45

## MENSURATION OF LINES AND SUPERFICIES.

The length and breadth of a square or rectangle being given, to find the area, . . . . 47
The area and one side of a square or rectangle being given, to find the other side, . . . 47
The base and perpendicular of a right-angled triangle being given, to find the hypotenuse, . . . . . . . . . . . . . . . . . . . . . . . . . . . . . . . . 48
The hypotenuse and one leg of a right-angled triangle being given, to find the other leg, . 49
The sum and difference of two numbers being given, to find the numbers, . . . . . . 50
The sum of two numbers and the difference of their squares being given, to find the numbers, . . . . . . . . . . . . . . . . . . . . . . . . . . . . . . . 50
One side of a right-angled triangle, and the sum of the hypotenuse and the other side being given, to find the hypotenuse and the other side, . . . . . . . . . . . . . 51
The relation of the three sides of a triangle to each other, applied to the measurement of distances, . . . . . . . . . . . . . . . . . . . . . . . . . . . . . . 52
To find the area of a right-angled triangle, . . . . . . . . . . . . . . . . . . . 54
To find the area of an equilateral and of an isosceles triangle, . . . . . . . . . . . 55
To find the area of any triangle, . . . . . . . . . . . . . . . . . . . . . . . . 56
To find the area of a rhombus and of a rhomboid, . . . . . . . . . . . . . . . . 57
To find the area of a trapezoid, . . . . . . . . . . . . . . . . . . . . . . . . 58
To find the area of a trapezium, . . . . . . . . . . . . . . . . . . . . . . . . 58
Similar Rectilinear Figures, . . . . . . . . . . . . . . . . . . . . . . . . . . 59
To find the area of any regular polygon, . . . . . . . . . . . . . . . . . . . . 60
Table of Regular Polygons, . . . . . . . . . . . . . . . . . . . . . . . . . . 61
To find the area of any rectilinear figure or polygon, . . . . . . . . . . . . . . . 62
Board or Lumber Measure, . . . . . . . . . . . . . . . . . . . . . . . . . . 63
The diameter of a circle being given, to find the circumference, . . . . . . . . . . 64
The circumference of a circle being given, to find the diameter, . . . . . . . . . . 66
The number of degrees in a circular arc, and the radius of the circle being given, to find the length of the arc, . . . . . . . . . . . . . . . . . . . . . . . . . 66
To find the area of a circle, . . . . . . . . . . . . . . . . . . . . . . . . . 67
To find the area of a circle when the diameter only is given, . . . . . . . . . . . 67
To find the diameter of a circle, the area being given, . . . . . . . . . . . . . . 68
To find the area of a semicircle, a quadrant, and a sextant, . . . . . . . . . . . . 68
To find the area of a sector, the radius and arc being given, . . . . . . . . . . . 69
To find the area of a sector, the radius and the angle at the center being given, . . . . 70
To find the side of a square which shall contain an area equal to a given circle, . . . 70
To find the side of an equilateral triangle inscribed in a given circle, . . . . . . . . 71
To find the side of a square inscribed in a given circle, . . . . . . . . . . . . . . 72
To find the side of an octagon inscribed in a given circle, . . . . . . . . . . . . 73
To find the area of an ellipse, . . . . . . . . . . . . . . . . . . . . . . . . 73
To find the diameter of a circle whose area shall be equal to that of a given ellipse, . . 74
To find the area of the space contained between the arcs of four equal adjacent circles, . 74
To find the area of the space contained between the arcs of three equal adjacent circles, . 75
To find the area of a circular ring, . . . . . . . . . . . . . . . . . . . . . . 76
Similar Curvilinear Figures, . . . . . . . . . . . . . . . . . . . . . . . . . 76
Practical Examples in the Mensuration of Lines and Superficies, . . . . . . . . . . 77

## MENSURATION OF SOLIDS.


To find the cubic contents of a prism, cube, parallelopiped, cylinder, or cylindroid, . 80
To find the cubic contents of a pyramid or a cone, . . . . . . . . . . . . . . . . 81
To find the hight of a pyramid or a cone, of which a given frustrum is a part, . . . . 82
To find the solidity of a frustrum of a pyramid or a cone, . . . . . . . . . . . . . 83
To find the superficies and the solidity of the regular solids, . . . . . . . . . . . . 84
Table of Regular Solids, . . . . . . . . . . . . . . . . . . . . . . . . . . . 84
To find the solidity of any irregular solid, . . . . . . . . . . . . . . . . . . . 85
To find the area of a sphere, . . . . . . . . . . . . . . . . . . . . . . . . . 86
To find the solidity of a sphere, . . . . . . . . . . . . . . . . . . . . . . . 86
Gauging, . . . . . . . . . . . . . . . . . . . . . . . . . . . . . . . . . 87
Timber Measure, . . . . . . . . . . . . . . . . . . . . . . . . . . . . . . 88
To find the contents of a four-sided stick of timber which tapers upon two opposite sides only, . . . . . . . . . . . . . . . . . . . . . . . . . . . . . . . 88
To find the contents of a stick of timber which tapers uniformly upon all sides, . . . 89
To find the number of cubic feet of timber any log will make when hewn square, . . 89
To find the number of feet of boards that can be sawn from any log of a given diameter, . 90
To find how many bushels of grain may be put into a bin of a given size, . . . . . . 92
Table for Boxes or Measures, Dry Measure, . . . . . . . . . . . . . . . . . . . 93
To find the side of the greatest cube that can be cut from any sphere, . . . . . . . 93
To find the weight of lead and iron balls, . . . . . . . . . . . . . . . . . . . 94
Practical Examples in the Mensuration of Solids, . . . . . . . . . . . . . . . . 94


## MECHANICAL POWERS.


The Lever, . . . . . . . . . . . . . . . . . . . . . . . . . . . . . . . . . 98
The Wheel and Axle, . . . . . . . . . . . . . . . . . . . . . . . . . . . . 101
The Pulley, . . . . . . . . . . . . . . . . . . . . . . . . . . . . . . . . 103
Smeaton's Pulley, . . . . . . . . . . . . . . . . . . . . . . . . . . . . . 105
The Inclined Plane, . . . . . . . . . . . . . . . . . . . . . . . . . . . . 106
The Wedge, . . . . . . . . . . . . . . . . . . . . . . . . . . . . . . . . 107
The Screw, . . . . . . . . . . . . . . . . . . . . . . . . . . . . . . . . 108
Friction, . . . . . . . . . . . . . . . . . . . . . . . . . . . . . . . . . 110
The Friction of Sliding Bodies, . . . . . . . . . . . . . . . . . . . . . . . 110
The Friction of Rolling Bodies, . . . . . . . . . . . . . . . . . . . . . . . 110
The Friction of the Axles of Wheels, . . . . . . . . . . . . . . . . . . . . . 111
General Remarks upon the Mechanical Powers, . . . . . . . . . . . . . . . . . 112


## MACHINERY.


Methods of Transmitting Motion, . . . . . . . . . . . . . . . . . . . . . . . 113
Spur, Crown, and Beveled or Conical Wheels, . . . . . . . . . . . . . . . . . . 114
The Universal Joint, . . . . . . . . . . . . . . . . . . . . . . . . . . . . 115
Teeth of Wheels, . . . . . . . . . . . . . . . . . . . . . . . . . . . . . . 115
Horse Power, . . . . . . . . . . . . . . . . . . . . . . . . . . . . . . . 116
Levers and Weighing Machines, . . . . . . . . . . . . . . . . . . . . . . . 117
Compound Lever, . . . . . . . . . . . . . . . . . . . . . . . . . . . . . . 117
The Balance, . . . . . . . . . . . . . . . . . . . . . . . . . . . . . . . 118
The Steelyard, . . . . . . . . . . . . . . . . . . . . . . . . . . . . . . 118
The Bent Lever Balance, . . . . . . . . . . . . . . . . . . . . . . . . . . 119
Wheel Work, . . . . . . . . . . . . . . . . . . . . . . . . . . . . . . . 119
White's Pulley, . . . . . . . . . . . . . . . . . . . . . . . . . . . . . . 121
The Crane, . . . . . . . . . . . . . . . . . . . . . . . . . . . . . . . . 122
Hunter's Screw, . . . . . . . . . . . . . . . . . . . . . . . . . . . . . . 123
The Endless Screw, . . . . . . . . . . . . . . . . . . . . . . . . . . . . 123
Pumps, . . . . . . . . . . . . . . . . . . . . . . . . . . . . . . . . . . 124
The Hydrostatic Press, . . . . . . . . . . . . . . . . . . . . . . . . . . . 125
The Tread-Mill, . . . . . . . . . . . . . . . . . . . . . . . . . . . . . . 126
Water Wheels, . . . . . . . . . . . . . . . . . . . . . . . . . . . . . . . 127
The Overshot Wheel, . . . . . . . . . . . . . . . . . . . . . . . . . . . . 127
The Undershot Wheel, . . . . . . . . . . . . . . . . . . . . . . . . . . . 127
The Breast Wheel, . . . . . . . . . . . . . . . . . . . . . . . . . . . . . 127

## ADAMS'S ARITHMETICAL SERIES.

**FOR SCHOOLS AND ACADEMIES.**

---

I. — PRIMARY ARITHMETIC, OR MENTAL OPERATIONS IN NUMBERS; being an introduction to the REVISED EDITION OF ADAMS'S NEW ARITHMETIC.

II. — ADAMS'S NEW ARITHMETIC, REVISED EDITION; in which the principles of operating by numbers are analytically explained and synthetically applied. Illustrated by copious examples.

III. — MENSURATION, MECHANICAL POWERS, AND MACHINERY. The principles of Mensuration analytically explained, and practically applied to the *measurement of lines, superficies and solids:* also, a philosophical explanation of the *simple mechanical powers,* and their application to *machinery.*

IV. — BOOK-KEEPING; containing a lucid explanation of the common method of BOOK-KEEPING BY SINGLE ENTRY; *a new, concise, and common-sense method of Book-keeping,* for farmers, mechanics, retailers, and professional men; methods of *keeping books by figures;* short methods of keeping accounts in a limited business; exercises for the pupil; and various forms necessary for the transaction of business. Accompanied with BLANK BOOKS, for the use of learners.

---

### ADVERTISEMENT.

The Primary Arithmetic, the Treatise on Mensuration, and the Book-keeping, have been mainly prepared, under my supervision, by Mr. J. HOMER FRENCH, of New York, who rendered important assistance in revising my New Arithmetic.

From my knowledge of his ability, and from a careful examination of the works, I can confidently say they meet my approbation.

DANIEL ADAMS.

*Keene, N. H. August,* 1848.

# MENSURATION.

---

## WEIGHTS AND MEASURES.

¶ **1.** *Measure* is that by which extent or dimension is ascertained, whether it be length, breadth, thickness, or amount. The process by which the extent or dimension is obtained is called *Measuring*, which consists in comparing the thing to be measured with some conventional standard.

*Weight* is the measure of the force by which any body, or a given portion of any substance, tends or gravitates to the earth. The process by which this measure is ascertained is called *Weighing*, which consists in comparing the thing to be measured with some conventional standard.

The United States government, after various unsuccessful attempts, at length succeeded, in the year 1834, in adopting a uniform standard of weights and measures, for the use of the custom-houses, and the other branches of business connected with the government. In the following tables the United States standards are given.

### Weights.

1. Troy Weight.

¶ **2.** *Troy Weight* is used where great accuracy is required, as in weighing gold, silver, and jewels. The denominations are pounds, ounces, pennyweights and grains.

**TABLE.**

| | | | | |
|---|---|---|---|---|
| 24 grains (grs.) | make | 1 pennyweight, | marked | pwt. |
| 20 pwts. | " | 1 ounce, | " | oz. |
| 12 oz. | " | 1 pound, | " | lb. |

---

¶ **1.** Measure. Measuring. Weight. Weighing. U. S. government standard weights and measures.

¶ **2.** Troy Weight. Denominations. Table.

¶ **3.** *The U. S. standard unit of weight* is the Troy pound of the mint, which is the same as the Imperial standard pound of Great Britain.

A cubic inch of distilled water in a vacuum, weighed by brass weights, also in a vacuum, at a temperature of 62° of Fahrenheit's thermometer, is equal to 252·724 grains, of which the standard Troy pound contains 5760. Consequently, a cubic inch of distilled water is $\frac{252724}{5760000}$ of a standard Troy pound. Hence, if the standard Troy pound be lost, destroyed, defaced, or otherwise injured, it may be restored of the same weight, by making a new standard, determined according to this ratio.

## II. Apothecaries' Weight.

¶ **4.** For the use of apothecaries and physicians, in compounding medicines, the Troy ounce is divided into drams, the drams into scruples, and the scruples into grains.

**TABLE.**

| | | | | |
|---|---|---|---|---|
| 20 grains (grs.) | make | 1 scruple, | marked | ℈ |
| 3 ℈ | " | 1 dram, | " | ʒ |
| 8 ʒ | " | 1 ounce, | " | ℥ |
| 12 ℥ | " | 1 pound, | " | ℔. |

Medicines are bought and sold by avoirdupois weight.

## III. Avoirdupois or Commercial Weight.

¶ **5.** *Avoirdupois Weight* (also called Commercial Weight) is employed in all the ordinary purposes of weighing. The denominations are tons, pounds, ounces, and drams.

**TABLE.**

| | | | | |
|---|---|---|---|---|
| 16 drams (drs.) | make | 1 ounce, | marked | oz. |
| 16 oz. | " | 1 pound, | " | lb. |
| 2000 lbs. | " | 1 ton, | " | T. |

Note. In the U. S. custom-house operations, in invoices of English goods, and of coal from the Pennsylvania mines, —

| | | | | |
|---|---|---|---|---|
| 28 lbs. | make | 1 quarter, | marked | qr. |
| 4 qrs. = 112 lbs. | " | 1 hundred weight, | " | cwt. |
| 20 cwt. = 2240 lbs. | " | 1 ton, | " | T. |

---

¶ **3.** U. S. standard unit of weight. How determined.

¶ **4.** Apothecaries' weight. Table.

¶ **5.** Avoirdupois or commercial weight. Denominations. Table. Note.

But in selling coal in cities, and in other transactions, unless otherwise stipulated, 2000 lbs. are called a ton.

The ton of 2240 lbs. is sometimes called the "long ton," and the ton of 2000 lbs. the "short ton."

**¶ 6.** *The U. S. avoirdupois pound* is determined from the standard Troy pound, and contains 7000 Troy grains, the Troy pound containing 5760.

The Troy pound is $\frac{5760}{7000}=\frac{144}{175}$, or nearly $\frac{14}{17}$, of an avoirdupois pound. But the Troy ounce contains ($\frac{5760}{12}=$) 480 grs., and the avoirdupois ounce ($\frac{7000}{16}=$) 437'5 grs. Troy. Therefore, the Troy ounce is greater than the avoirdupois ounce in the ratio of 480 to 437'5 = 4800 to 4735 = 192 to 175.

**¶ 7.** *The standard pound of the State of New York* is the pound avoirdupois, which is defined, by declaring that a cubic foot of pure water, at its maximum density, (39'83° Fahrenheit,) weighs 62'5 pounds, or 1000 ounces, using brass weights, at the mean pressure of the atmosphere at the level of the sea, (i. e., the barometer being at 30 inches.) Therefore, the standard pound of the State of N. Y. is the weight of 27'648 cubic inches of distilled water, weighed in air, the temperature being 39'83° Fahrenheit, and the barometer at 30 inches.

---

## Measures of Extension.

### I. 1. Linear Measure.

**¶ 8.** *Linear Measure* (also called Long Measure) is the measure of lines; it is used when only one dimension is considered, which may be length, breadth, or thickness.

The usual dimensions are miles, furlongs, rods, yards, feet, and inches.

**TABLE.**

| | | | | | |
|---|---|---|---|---|---|
| 12 inches (in.) | make | 1 foot, | marked | ft. |
| 3 ft. | " | 1 yard, | " | yd. |
| $5\frac{1}{2}$ = 5'5 yds., or $16\frac{1}{2}$ = 16'5 ft., | " | 1 rod, | " | rd. |
| 40 rds. | " | 1 furlong, | " | fur. |
| 8 fur., or 320 rds., | " | 1 mile, | " | mi. |

---

¶ 6. U. S. avoirdupois pound. Ratio of the Troy pound to the avoirdupois pound. — of the Troy ounce to the avoirdupois ounce.

¶ 7. N. Y. standard pound. Maximum density and mean pressure of the atmosphere.

¶ 8. Linear measure. Denominations. Table.

## 2. Cloth Measure.

**¶ 9.** *Cloth Measure* is used in measuring cloth and other goods sold by the yard in length, without regard to width.

**TABLE.**

| | | | | |
|---|---|---|---|---|
| $2\frac{1}{4}$ = 2·25 inches (in.) | make | 1 nail, | marked | na. |
| 4 na., or 9 in., | " | 1 quarter, | " | qr. |
| 4 qrs., or 36 in., | " | 1 yard, | " | yd. |

## 3. Linear Chain Measure.

**¶ 10.** *The Surveyor's*, or what is called *Gunter's Chain*, is generally used in measuring long distances, and in surveying land.

It is 4 rods, or 66 feet, in length, and consists of 100 links.

**TABLE.**

| | | | | | |
|---|---|---|---|---|---|
| 7·92 | inches (in.) | make | 1 link, | marked | l. |
| 25 | l. | " | 1 rod, | " | rd. |
| 4 | rds., 66 ft., or 100 l., | " | 1 chain, | " | C. |
| 80 | C. | " | 1 mile, | " | mi. |

## 4. Duodecimal Linear Measure.

**¶ 11.** *Duodecimals* are fractions of a foot. The denominations are fourths, thirds, seconds, primes or inches, and feet.

**TABLE.**

| | | | | |
|---|---|---|---|---|
| 12 fourths ('''') | make | 1 third, | marked | ''' |
| 12 ''' | " | 1 second, | " | '' |
| 12 '' | " | 1 prime, or inch, | " | ' |
| 12 ' | " | 1 foot, | " | ft. |

Note. The marks, ', '', ''', '''', &c., which distinguish the different parts, are called the *indices of the parts* or denominations. [See Adams's Revised Arithmetic, ¶¶ 203 and 204.]

## ¶ 12. 5. Miscellaneous Linear Measures.

| | | | |
|---|---|---|---|
| 6 points | make | 1 line, | used in measuring the length of the rods of clock pendulums. |
| 12 lines | " | 1 inch, | |
| 4 inches | " | 1 hand, | used in measuring the hight of horses. |

---

¶ **9.** Cloth measure. Table.

¶ **10.** Linear chain measure. Gunter's chain. Table.

¶ **11.** Duodecimal linear measure. Denominations. Table. Note.

¶ **12.** Miscellaneous linear measures.

6 feet make 1 fathom, used in measuring depths at sea.
18 inches " 1 cubit.
21‘888 in. " 1 sacred cubit.

$69\frac{1}{6}$ common miles make 1 degree, or °, { on the equatorial circum. of the earth.

3 geographical miles make 1 league, (L.) { used in measuring distances at sea.

60 geographical miles make 1 degree of latitude.

**¶ 13.** *The U. S. standard of measures of extension,* whether linear, superficial, or solid, is the yard of 3 feet, or 36 inches, and is the same as the Imperial standard yard of Great Britain. The standard yard is made of brass, and is laid off from a scale made by Troughton, (a celebrated English artist,) the brass being at the temperature of 62° Fahrenheit's thermometer.

The standard yard, when compared with the length of the rod of a pendulum vibrating seconds of mean time in the latitude of London, in a vacuum at the level of the sea, is found to be in the ratio of 36 inches to 39‘1393. Hence, if the standard yard be lost or destroyed, it may be restored, by making it $\frac{360000}{391393}$ of the length of the rod of a pendulum vibrating seconds under the above described circumstances.

**¶ 14.** *The standard yard of the State of N. Y.* is a brass rod, which bears to the length of the rod of a pendulum vibrating seconds in a vacuum, in Columbia College, the relation of 1000000 to 1086141, the brass being at 32° Fahrenheit.

## II. 1. Square Measure.

**¶ 15.** *Square Measure* is used in measuring all things wherein length and breadth are considered; such as land, flooring, roofing, painting, plastering, &c.

The denominations are miles, acres, roods, rods or poles, yards, feet, and inches.

**TABLE.**

144 square inches (sq. in.) make 1 square foot, marked sq. ft.
9 sq. ft. " 1 square yard, " sq. yd.

---

¶ **13.** U. S. standard of measures of extension. Material of standard yard. Temperature. Comparison of standard yard with the length of the rod of a pendulum. Standard yard, if lost, may be restored.

¶ **14.** N. Y. standard yard.

¶ **15.** Square measure. Denominations. Table.

| | | | |
|---|---|---|---|
| $30\frac{1}{4}$ = 30·25 sq. yds., or $272\frac{1}{4}$ = 272·25 sq. ft. | make | 1 square rod, or pole, | marked sq. rd. or P. |
| 40 sq. rds. | " | 1 rood, | marked R. |
| 4 R., or 160 sq. rds., | " | 1 acre, | " A. |
| 640 A. | " | 1 square mile, | " M. |

### 2. Square Chain Measure.

**¶ 16.** The dimensions of land are generally taken by Gunter's chain, and are estimated by the following

**TABLE.**

| | | | |
|---|---|---|---|
| 625 square links (sq. l.) | make | 1 square rod, or pole, | marked sq. rd. or P. |
| 16 P. | " | 1 square chain, marked | sq. C. |
| 10 sq. C. | " | 1 acre, | " A. |
| 640 A., or 6400 sq. C., | " | 1 square mile, | " M. |

### 3. Artificers' Superficial Measure.

**¶ 17.** Artificers estimate their work as follows:

1. Glazing, and stone-cutting, by the square foot.
2. Painting, plastering, paving, ceiling, and paper-hanging, by the square yard.
3. Flooring, partitioning, roofing, slating and tiling, by the square of 100 feet.
4. Brick-laying, by the thousand bricks; also, by the square yard, and the square of 100 feet.

Note 1. In estimating the painting of mouldings, cornices, &c., the measuring line is carried into all the mouldings and cornices.

Note 2. In estimating brick-laying by the square yard, or by the square of 100 feet, the work is understood to be $1\frac{1}{2}$ bricks, or 12 inches thick. If it be of any other thickness, it must be reduced to that of $1\frac{1}{2}$ bricks, before estimating the brick-laying.

## III. Cubic or Solid Measure.

**¶ 18.** *Cubic or Solid Measure* is used in measuring things that have length, breadth, and thickness; such as timber, wood, earth, stone, &c.

The denominations are cords, tons, yards, feet, and inches.

---

**¶ 16.** Square chain measure. Table.

**¶ 17.** Artificers' superficial measure. Note 1. Note 2.

**¶ 18.** Cubic or solid measure. Denominations. Table. A cubic yard of earth. A cubic ton. A cord of wood. A cord foot. A perch of stone, or masonry. Note 1. Note 2.

TABLE.

| | | | | |
|---|---|---|---|---|
| 1728 cubic inches (cu. in.) | make | 1 cubic foot, | marked | cu. ft. |
| 27 cu. ft. | " | 1 cubic yard, | " | cu. yd. |
| 50 cu. ft. of round timber, or<br>40 cu. ft. of hewn timber, | " | 1 ton, | " | T. |
| 42 cu. ft. | " | 1 ton of shipping, | " | T. |
| 16 cu. ft. | " | 1 cord foot, or<br>1 foot of wood, | " | C. ft. |
| 8 C. ft., or 128 cu. ft., | " | 1 cord of wood, | " | C. |
| $24\frac{3}{4}=24{\cdot}75$ cu. ft. | " | 1 perch of stone, | " | Pch. |

A cubic yard of earth is called a load.

A cubic ton is used for estimating the cartage and transportation of timber. A ton of round timber is such a quantity (about 50 feet) as will make 40 feet when hewn square.

A pile of wood 8 feet long, 4 feet wide, and 4 feet high, contains 1 cord; and a cord foot is 1 foot in length of such a pile.

A perch of stone or of masonry is $16\frac{1}{2}$ feet long, $1\frac{1}{2}$ feet wide, and 1 foot high. If any wall be $1\frac{1}{2}$ feet thick, its contents in perches will be equal to the number of times $16\frac{1}{2}$ square feet are contained in the superficial contents of the wall expressed in feet. If the wall be of any other thickness, the number of perches it contains will be found by dividing its cubic contents by the cubic contents of a perch.

NOTE 1. Joiners, brick-layers, and masons make no allowance for windows, doors, &c. Brick-layers and masons, in estimating their work by cubic measure, make no allowance for the corners to the walls of houses, cellars, &c., but estimate their work by the girt, that is, the length of the wall on the outside.

NOTE 2. Engineers, in making estimates for excavations and embankments, take the dimensions with a line, or measure, divided into feet and decimals of a foot. The estimates are made in feet and decimals, and are then reduced to cubic yards.

---

## Measures of Capacity.

### I. WINE MEASURE.

**¶ 19.** *Wine Measure* is used in measuring all liquids except ale, beer and milk.

The denominations are gallons, quarts, pints, and gills.

---

¶ **19.** Wine measure. Denominations. Table. Note.

TABLE.

| | | | | |
|---|---|---|---|---|
| 4 gills (gi.) | make | 1 pint, | marked | pt. |
| 2 pts. | " | 1 quart, | " | qt. |
| 4 qts. | " | 1 gallon, | " | gal. |

NOTE. The following denominations are also sometimes used in this measure:—

| | | | | |
|---|---|---|---|---|
| 31½ gal. | make | 1 barrel, | marked | bar. |
| 42 gal. | " | 1 tierce, | " | tier. |
| 63 gal., or 2 bar., | " | 1 hogshead, | " | hhd. |
| 2 hhds. | " | 1 pipe, or butt, | " | P. |
| 2 P., or 4 hhds., | " | 1 tun, | " | T. |

But the tierce, hogshead, puncheon, pipe, butt, and tun, used for liquids, are so vague and variable in their contents, that they are to be considered rather as the names of casks than as expressing any fixed or definite measures. Such vessels are usually gauged, and have their contents marked on them.

**¶ 20.** *The U. S. standard of liquid measure* is the old English wine gallon, of 231 cubic inches, equal to 8·339 pounds avoirdupois of distilled water, at the maximum density 39·83° Fahrenheit, the barometer at 30 inches.

**¶ 21.** *The standard of liquid measure in the State of New York* is the wine gallon, which the legislature have defined to be equal to 8 pounds of pure water at its maximum density. Hence the N. Y. wine gallon contains 221·184 cubic inches.

## II. BEER MEASURE.

**¶ 22.** *Beer Measure* is used in measuring beer, ale, and milk.

The denominations are hogsheads, barrels, gallons, quarts, and pints.

TABLE.

| | | | | |
|---|---|---|---|---|
| 2 pints (pts.) | make | 1 quart, | marked | qt. |
| 4 qts. | " | 1 gallon, | " | gal. |
| 36 gal. | " | 1 barrel, | " | bar. |
| 54 gal., or 1½ bar., | " | 1 hogshead, | " | hhd. |

The gallon Beer Measure contains 282 cubic inches.

Beer Measure is retained in use only by custom. In many places its use is entirely discarded.

---

**¶ 20.** U. S. standard of liquid measure.

**¶ 21.** N. Y. standard of liquid measure.

**¶ 22.** Beer measure. Denominations. Table. Beer gallon. Authority for using this measure.

## III. Dry Measure.

**¶ 23.** *Dry Measure* is used in measuring grain, fruit, roots, salt, coal, &c.

The denominations are bushels, pecks, quarts and pints.

**TABLE.**

| | | | |
|---|---|---|---|
| 2 pints (pts.) | make | 1 quart, | marked qt. |
| 8 qts. | " | 1 peck, | " pk. |
| 4 pks. | " | 1 bushel, | " bu., or bush. |

The quarter of 8 bushels is an English measure for grain. The chaldron of 36 bushels is sometimes used in measuring charcoal.

**¶ 24.** *The U. S. standard of dry measure* is the British Winchester bushel, which is 18½ inches in diameter, and 8 inches deep, and contains 2150'4 cubic inches, equal to 77'6274 pounds of distilled water, at the maximum density. A gallon dry measure contains 268'8 cubic inches.

**¶ 25.** *The standard bushel of the State of N. Y.* contains 80 pounds, or 2211'84 cubic inches of pure water at its maximum density; the gallon contains 10 pounds, or 276'48 cubic inches.

Note. *The Imperial gallon of Great Britain*, for all liquids and dry substances, contains 277'274 cubic inches, or 10 pounds avoirdupois weight of distilled water weighed in air, at 62° Fahrenheit, the barometer at 30 inches.

*The Imperial standard bushel* contains 2218'192 cubic inches, or 80 pounds of distilled water, weighed in the manner above described.

---

**¶ 23.** Dry measure. Denominations. Table. Quarter. Chaldron.

**¶ 24.** U. S. standard of dry measure. Dry gallon.

**¶ 25.** N. Y. standard bushel. N. Y. dry gallon. Imperial gallon and bushel of Great Britain.

## ¶ 26. Standard Road Measures of different Countries.

| Countries. | Measures. | Yards. | English mile. |
|---|---|---|---|
| China, . . . . . . . . | Li, . . . . . . . | 629 | ‘357 |
| Corsica, France, French Guiana, Netherlands, Modern Greece, and French West Indies, . . . . . . . | Mile, . . . . . . | 1093‘633 | ‘621 |
| Russia, . . . . . . . . | Werst, . . . . . . | 1167 | ‘663 |
| Sicily, . . . . . . . . | Mile, . . . . . . | 1405‘555 | ‘798 |
| Ancient Greece, . . . . | " | 1611‘666 | ‘915 |
| Ancient Rome, . . . . | " | 1614 | ‘917 |
| Modern Rome . . . . | " | 1628 | ‘925 |
| Australia, Canada, Ceylon, Gibraltar, Great Britain, British Guiana, N. Brunswick, Newfoundland, Nova Scotia, United States, and British West Indies, . . | " | 1760 | 1‘000 |
| Italy, . . . . . . . . | Old common mile, . | 1766‘333 | 1‘003 |
| Tuscany, . . . . . . . | Mile, . . . . . | 1808 | 1‘027 |
| Lombardy, . . . . . . . | " | 1813‘432 | 1‘030 |
| Turkey, . . . . . . . . | Barri, . . . . . . | 1826 | 1‘037 |
| Scotland, . . . . . . . | Ancient mile, . . . | 1984 | 1‘127 |
| Bengal, (India,) . . . . | League, . . . . . | 2000 | 1‘136 |
| Naples, . . . . . . . . | Mile, . . . . . . | 2018 | 1‘146 |
| Australia, Canada, Ceylon, Gibraltar, Great Britain, British Guiana, N. Brunswick, Newfoundland, Nova Scotia, United States, and British West Indies, . . | Geographical mile, . | 2025 | 1‘150 |
| Arabia, . . . . . . . . | Mile, . . . . . . | 2148 | 1‘220 |
| Ireland, . . . . . . . . | " | 2240 | 1‘272 |
| Brazil, Madeira, and Portugal, | " | 2253 | 1‘280 |
| Ancient Jewish, . . . . | " | 2432 | 1‘381 |
| Piedmont, . . . . . . . | " | 2640 | 1‘500 |
| Sardinia, . . . . . . . | " | 2697 | 1‘532 |
| Corsica, France, French W. Indies, Netherlands, French Guiana, and Greece, . . | Post league, . . . | 4264 | 2‘422 |
| Argentine Rep., Buenos Ayres, Chili, Cuba, Mexico, New Granada, Peru, Philippine Islands, and Spain, . | Judicial league, . . | 4635 | 2‘633 |
| Corsica, France, Greece, French Guiana, Netherlands, and French West Indies, . . . . . . . | Common league, . . | 4861 | 2‘761 |

| Countries. | Measures. | Yards. | English mile. |
|---|---|---|---|
| Australia, Canada, Ceylon, Corsica, France, Gibraltar, G. Britain, Greece, British Guiana, French Guiana, N. Brunswick, Newfoundland, Nova Scotia, Netherlands, United States, British W. Indies, and French West Indies, . . . . | Marine or sea league, | 6075 | 3'451 |
| Persia, . . . . . . . . | Parasang, . . . . | 6086 | 3'457 |
| Belgium, Dutch Guiana, Holland, and Dutch W. Indies, | Mile, . . . . . | 6395 | 3'633 |
| Brazil, Madeira, & Portugal, | League, . . . . | 6760 | 3'840 |
| Germany, . . . . . . . | Short mile, . . . | 6859 | 3'897 |
| Flanders, . . . . . . . | League, . . . . | 6864 | 3'900 |
| Argentine Republic, Buenos Ayres, Chili, Cuba, Mexico, New Granada, Peru, Philippine Islands, and Spain, . . . . . . . | Common league, . . | 7416 | 4'213 |
| Germany, . . . . . . . | Geographical mile, . | 8100 | 4'602 |
| Poland, . . . . . . . . | Long mile, . . . | 8100 | 4'602 |
| Wertemburg, . . . . . . | Mile, . . . . . . | 8100 | 4'602 |
| Prussia, . . . . . . . | " | 8237 | 4'680 |
| Denmark, Hamburg, Norway, and Danish West Indies, | " | 8244 | 4'684 |
| Vienna, . . . . . . . . | Post mile, . . . . | 8296 | 4'713 |
| Dantzic, . . . . . . . | Mile, . . . . . . | 8475 | 4'815 |
| Hungary, . . . . . . . | " | 9113 | 5'177 |
| Switzerland, . . . . . . | League, . . . . | 9153 | 5'200 |
| Saxony, . . . . . . . . | Mile, . . . . . . | 9914 | 5'632 |
| Germany, . . . . . . . | Long mile, . . . | 10126 | 5'753 |
| Bohemia, . . . . . . . | League, . . . . | 10137 | 5'759 |
| Hanover, . . . . . . . | Mile, . . . . . . | 11559 | 6'567 |
| Sweden, and Swedish West Indies, . . . . . . . | League, . . . . | 11700 | 6'647 |

# DEFINITIONS.

¶ **27.** *A Line* is that which has length, without breadth or thickness. Thus, *AB* and *CD* are lines.

*A Superficies or Surface* is a figure that has length and breadth, without thickness. Thus *ABCD* is a superficies or surface.

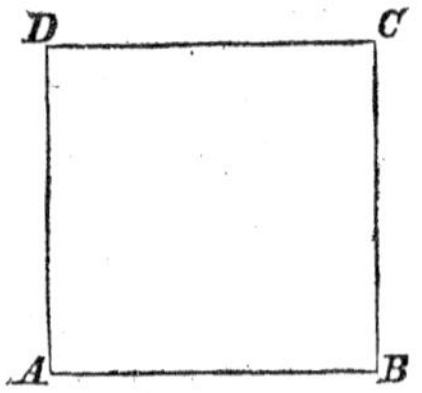

*A Solid* is a figure that has length, breadth, and thickness. Thus, *ABC DEFGH* is a solid.

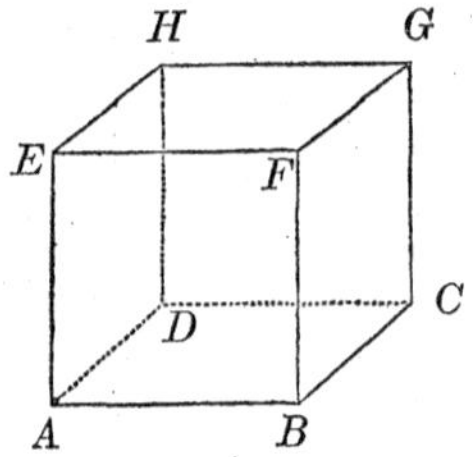

*Magnitude* is that which has one or more of the three dimensions, length, breadth, and thickness.

*Mensuration* is the art of measuring lines, surfaces, and solids. It is divided into three general classes, *Lineal, Superficial, and Solid measures.*

*Lineal Measure* is the measure of length.

*Superficial Measure* is the measure of length and breadth, or of surface.

*Solid Measure* is the measure of length, breadth, and thickness, or of solidity.

---

¶ **27.** Topic. A line. A superficies or surface. A solid. Magnitude. Mensuration. Lineal measure. Superficial measure. Solid measure. Note. Classification of mensuration. Geometry. Geometrical figures.

NOTE. Superficial and solid measures are used only in estimating the superficial and solid contents of figures, the dimensions of the figures always being taken in lineal measure.

Mensuration in its various forms is classed under that branch of mathematics called *Geometry*. Hence,

*Geometry* is the science of magnitude in general. The figures generally considered in Mensuration are called *Geometrical Figures*.

---

## GEOMETRICAL DEFINITIONS.

### LINES AND ANGLES.

**¶ 28.** 1. *A Point* is that which has position, but not magnitude. Thus, *A* is a point.

2. *A Line* may be either right (straight) or curved.

3. *A Right Line* is the shortest distance that can be drawn between two points. Thus, *AB* is a right line.

4. *A Curve Line* is that which is neither a right line nor composed of right lines. Thus, *AB* and *CD* are curve lines.

NOTE 1. A right line is commonly called a *line*, and a curve line a *curve*.

5. *Parallel Lines* are those which run in the same direction, at an equal distance from each other, and never meet. Thus, the lines *AB* and *CD* are parallel to each other.

6. *Parallel or Concentric Curves* are those which are equally distant from each other at every point. Thus, the curves *AB* and *CD* are parallel to each other.

7. *A Horizontal Line* is a line drawn parallel to the horizon. Thus, the line *AB* is horizontal.

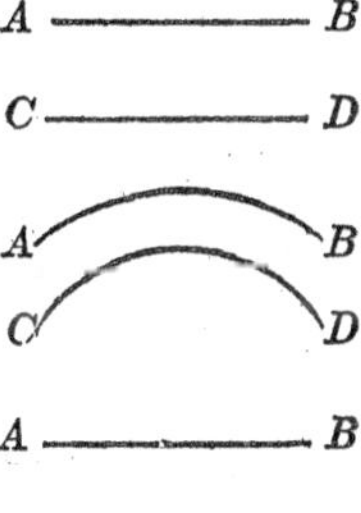


8. *A Vertical Line* is one which extends in a right line from some point towards the center of the earth. Thus, the line *CD* is vertical.

---

**¶ 28.** Topic. A point. A line. A right line. A curve line. Note 1. Parallel lines. Parallel or concentric curves. A horizontal line. A vertical

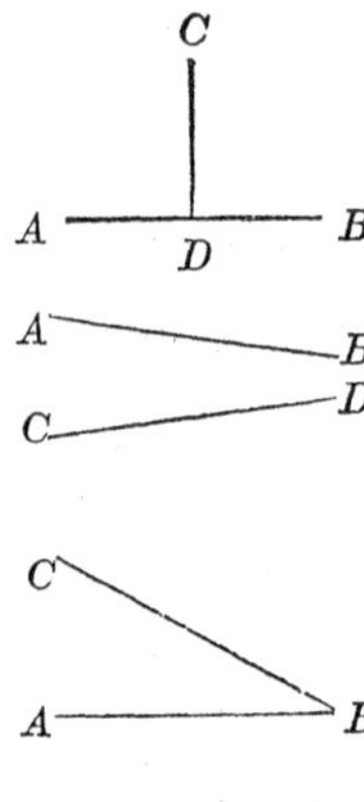


9. One line is said to be *Perpendicular* to another, when it so stands upon the other as to incline to neither side. Thus, the line *CD* is perpendicular to the line *AB*.

10. *Oblique Lines* are those which continually approach each other. Thus, the lines *AB* and *CD* are oblique.

11. *An Angle* is the space comprised between two lines that meet in a point. The point of meeting is the *Vertex* of the angle, and the lines containing the angle are its *Sides* or *Legs*. Thus, the space comprised between the lines *AB* and *CB* is an angle; the point *B* is its vertex; and the lines *AB* and *CB* are its sides or legs.

NOTE 2. An angle is generally read by placing the letter at the vertex in the middle; thus, the angle *A B C*. Or, the letter at the vertex only may be named; thus, the angle *B*.

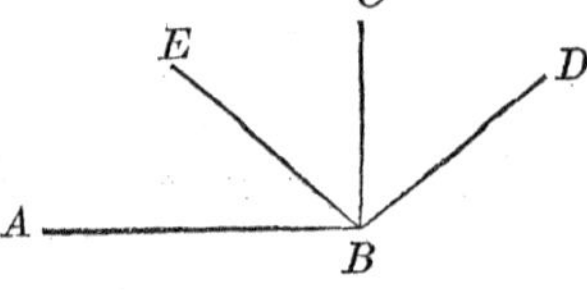


12. *A Right Angle* is one formed by one right line falling on another perpendicularly. Thus, *ABC* is a right angle.

13. *An Obtuse Angle* is greater than a right angle. Thus, *ABD* is an obtuse angle.

14. *An Acute Angle* is less than a right angle. Thus, *ABE* is an acute angle.

NOTE 3. Obtuse and acute angles are also called *Oblique Angles*.

15. *A Rectilinear or Right-Lined Angle* is formed by two lines. It may be right, obtuse, or acute. Thus, *ABC*, *ABD*, and *ABE*, are rectilinear angles.

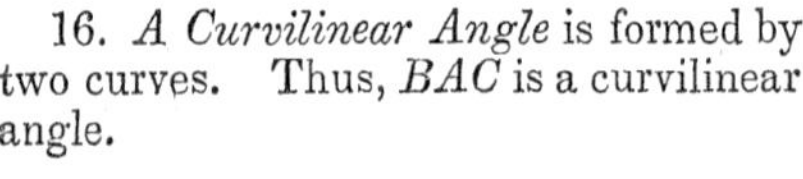

16. *A Curvilinear Angle* is formed by two curves. Thus, *BAC* is a curvilinear angle.

line. Lines perpendicular to each other. Oblique lines. An angle. Its vertex. Its sides or legs. Note 2. A right angle. An obtuse angle. An acute angle. Note 3. A rectilinear or right-lined angle. A curvilinear angle. A mixed angle. Adjacent or contiguous angles.

17. *A Mixed Angle* is formed by a line and a curve. Thus, *ABC* is a mixed angle.

18. *Adjacent or Contiguous Angles* are such as have one leg common to both angles. Thus, the angles *ABD* and *DBC* are contiguous.

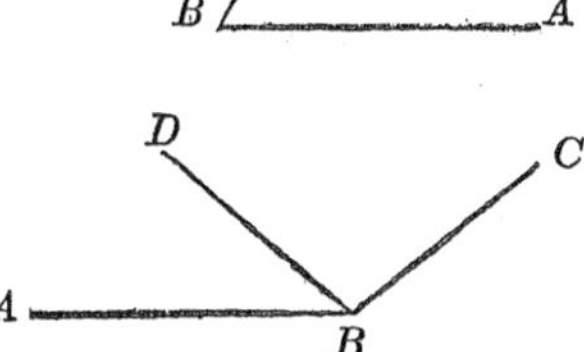


---

PLANE FIGURES.

**¶ 29.** 1. *Plane Figures* are even or level surfaces, bounded on all sides by lines or curves.

2. *Rectilinear Plane Figures* are planes bounded by lines.

3. *Curvilinear Plane Figures* are planes bounded by curves.

4. *Mixtilinear Plane Figures* are planes bounded by lines and curves.

RECTILINEAR PLANE FIGURES.

**¶ 30.** 1. Rectilinear plane figures are called *Polygons.*

2. *A Regular Polygon* is one whose sides are all equal.

3. *An Irregular Polygon* is one whose sides are unequal.

4. *The Perimeter* of a polygon is the sum of all its sides, or the distance round it.

5. *Similar Rectilinear Figures* are such as have their several angles respectively equal each to each, and their sides about the equal angles proportional.

6. *A Triangle* is a polygon of three sides. Thus, *ABC* is a triangle.

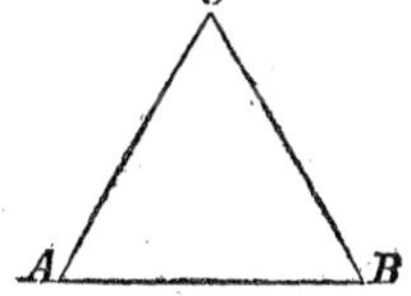


---

¶ **29.** Topic. Plane figures. Rectilinear plane figures. Curvilinear plane figures. Mixtilinear plane figures.

¶ **30.** Topic. Polygons. A regular polygon. An irregular polygon. Perimeter of a polygon. Similar rectilinear figures. Triangle. Quadri-

7. *A Quadrilateral* is a polygon of four sides. Thus, *ABCD* is a quadrilateral.

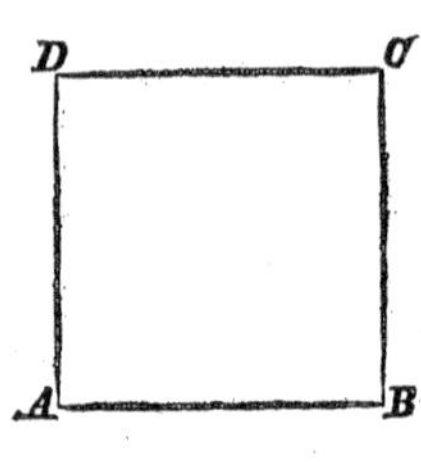


8. *A Pentagon* is a polygon of five sides. Thus, *ABCDE* is a pentagon.

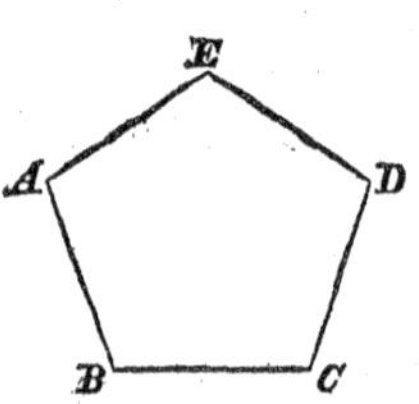


9. *A Hexagon* is a polygon of six sides. Thus, *ABCDEF* is a hexagon.

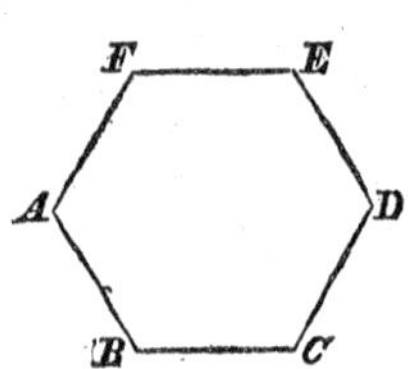


10. *A Heptagon* is a polygon of seven sides. Thus, *ABCDEFG* is a heptagon.

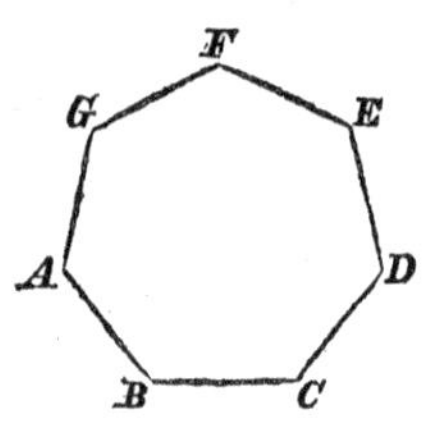


11. *An Octagon* is a polygon of eight sides. Thus, *ABCDEFGH* is an octagon.

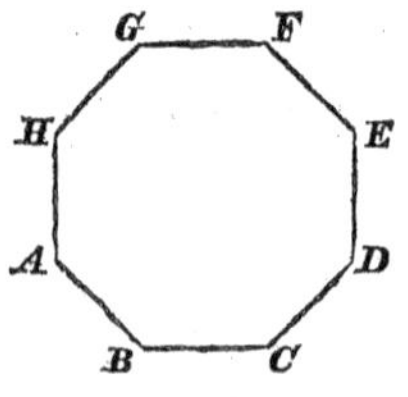


---

lateral. Pentagon. Hexagon. Heptagon. Octagon. Nonagon. Decagon.

12. *A Nonagon* is a polygon of nine sides.
13. *A Decagon* is a polygon of ten sides.
14. *An Undecagon* is a polygon of eleven sides.
15. *A Dodecagon* is a polygon of twelve sides.

Triangles are distinguished as *Right-angled*, *Obtuse-angled*, *Acute-angled*, *Equilateral*, *Isosceles*, and *Scalene*.

16. *A Right-angled Triangle* has one right angle. Thus, *ABC* is a right-angled triangle.

NOTE 1. A right-angled triangle is also called a *Rectangular Triangle*.

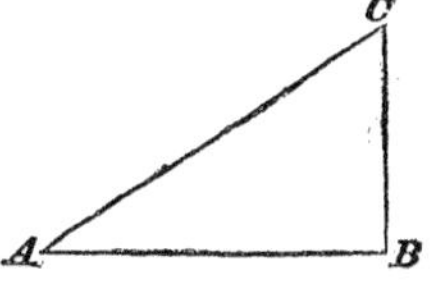

17. *An Obtuse-angled Triangle* has one obtuse angle. Thus, *ABC* is an obtuse-angled triangle.

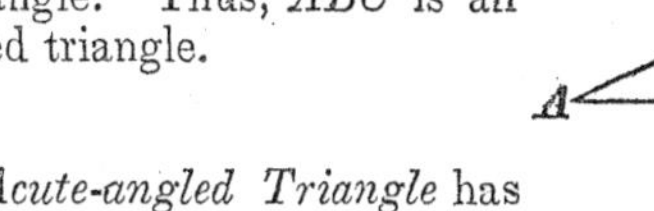

18. *An Acute-angled Triangle* has all the three angles acute. Thus, *ABC* is an acute-angled triangle.

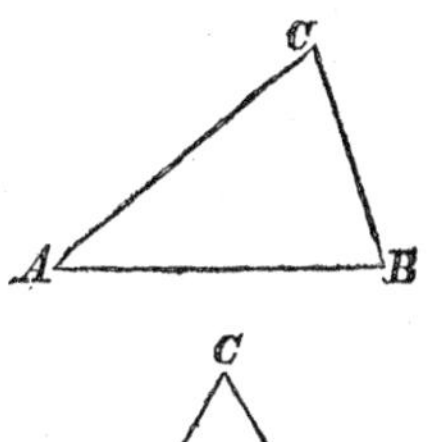

NOTE 2. Obtuse-angled and acute-angled triangles are also called *Oblique-angled Triangles*.

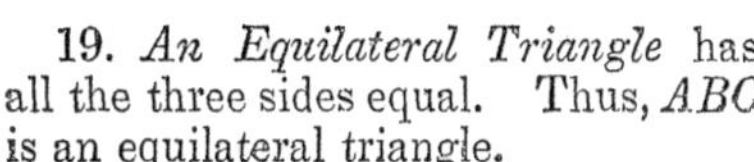

19. *An Equilateral Triangle* has all the three sides equal. Thus, *ABC* is an equilateral triangle.

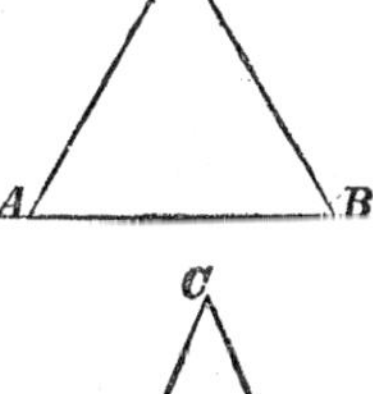

20. *An Isosceles Triangle* has only two of its sides equal. Thus, *ABC* is an isosceles triangle.

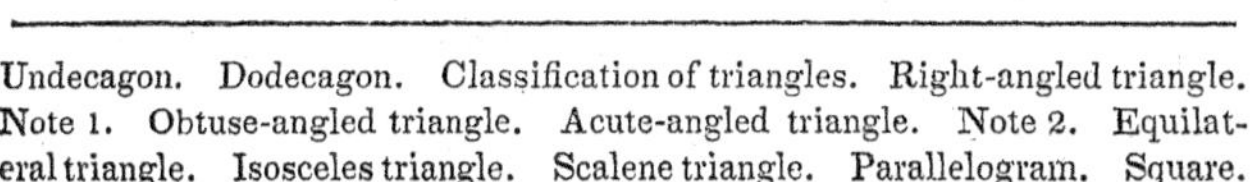

---

Undecagon. Dodecagon. Classification of triangles. Right-angled triangle. Note 1. Obtuse-angled triangle. Acute-angled triangle. Note 2. Equilateral triangle. Isosceles triangle. Scalene triangle. Parallelogram. Square.

3

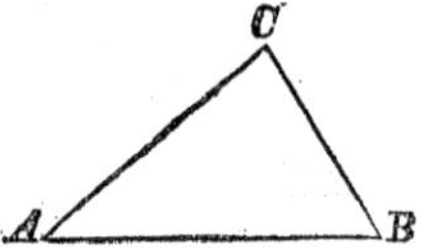

21. *A Scalene Triangle* has all the three sides unequal. Thus, *ABC* is a scalene triangle.

22. *A Parallelogram* is a right-lined figure, whose opposite sides are parallel, and consequently equal.

23. *A Square* is a figure having four equal sides and four right angles. It is a parallelogram whose sides are all equal, and whose angles are all right angles. Thus, *ABCD* is a square.

24. *A Rectangle* is a right-angled parallelogram, whose length exceeds its breadth. Thus, *ABCD* is a rectangle.

NOTE 3. The *areas* of rectangles and squares are sometimes called rectangles.

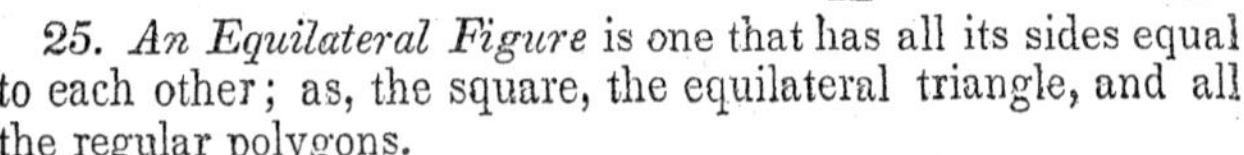

25. *An Equilateral Figure* is one that has all its sides equal to each other; as, the square, the equilateral triangle, and all the regular polygons.

26. *An Equiangular Figure* is one that has all its angles equal to each other; as all the regular polygons.

27. *A Quadrilateral Figure* is one contained by four right lines; as, the square, the rectangle, &c.

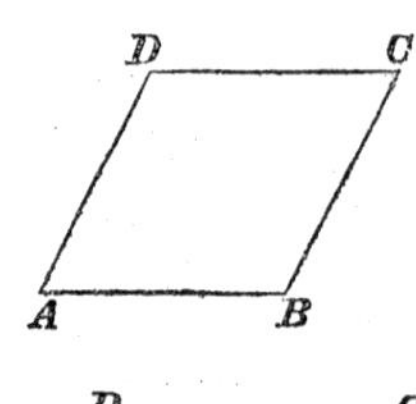

28. *A Rhombus* or *Rhomb* is an oblique-angled equilateral parallelogram. It is a quadrilateral whose sides are equal, and the opposite sides parallel, but the angles unequal, two being obtuse and two acute. Thus, *ABCD* is a rhombus.

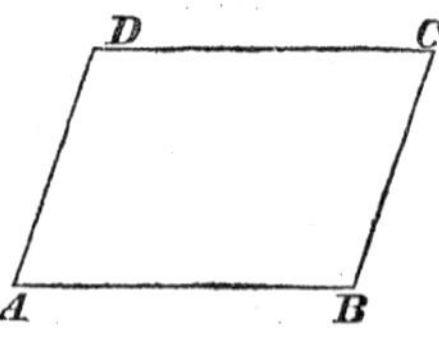

29. *A Rhomboid* is an oblique-angled parallelogram. It is a quadrilateral whose opposite sides and angles are equal, but which are neither equilateral nor equiangular. Thus, *ABCD* is a rhomboid.

---

Rectangle. Note 3. An equilateral figure. An equiangular figure. A

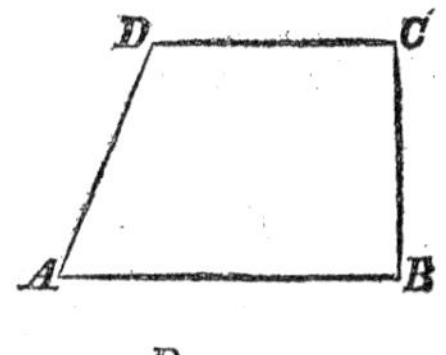


30. *A Trapezoid* is a quadrilateral which has two of the opposite sides parallel. Thus, *ABCD* is a trapezoid.

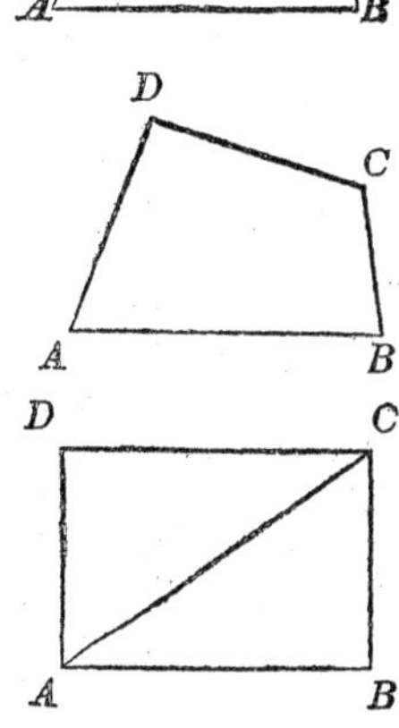


31. *A Trapezium* is a quadrilateral which has not two sides parallel. Thus, *ABCD* is a trapezium.

32. *A Diagonal* is a line drawn through a figure, joining two opposite angles. Thus, *AC* is the diagonal of the rectangle *ABCD*.

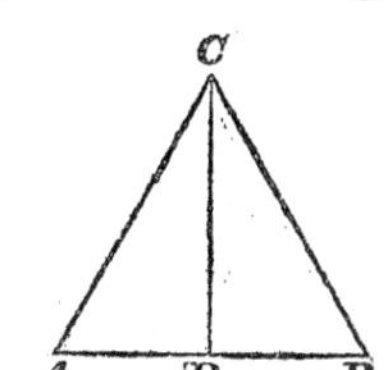


33. *The Apex* of a figure is its highest point. Thus, *C* is the apex of the triangle *ABC*.

34. *The Altitude* of a figure is the perpendicular hight of its apex above its base. Thus, *DC* is the altitude of the triangle *ABC*.

## Curvilinear and Mixtilinear Plane Figures.

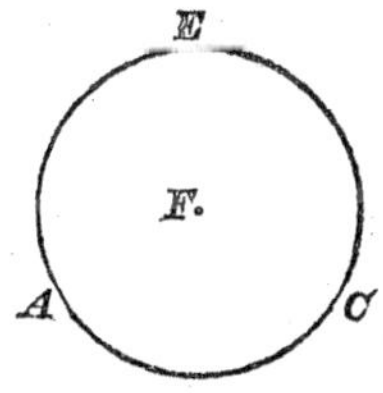


**¶ 31.** 1. *A Circle* is a plane figure comprehended by a single curve, called its *Circumference* or *Periphery*, every part of which is equally distant from a point called the *Center*. Thus, the space inclosed by the curve *ACE* is a circle, the curve is the circumference or periphery, and the point *F* is the center.

---

quadrilateral figure. Rhombus, or rhomb. Rhomboid. Trapezoid. Trapezium. Diagonal. Apex. Altitude.

NOTE 1. The circumference of a circle, for the sake of brevity, is frequently called a circle.

2. *The Diameter* of a circle is a line passing through the center, and terminating at each end in the circumference. It divides the circle into two equal parts, called *Semi-circles*. Thus, *AD* is the diameter of the circle *ABDE*, and *ABD* and *AED* are semi-circles.

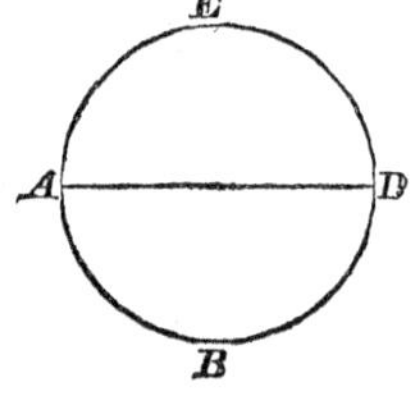


3. *The Radius* of a circle is a line extending from the center to the periphery. It is the semi-diameter. Two or more such lines are called *Radii*. All radii of a circle are equal to each other. Thus, *AF*, *CF*, *DF*, and *EF*, are radii of the circle *ACDE*, and are all equal to each other.

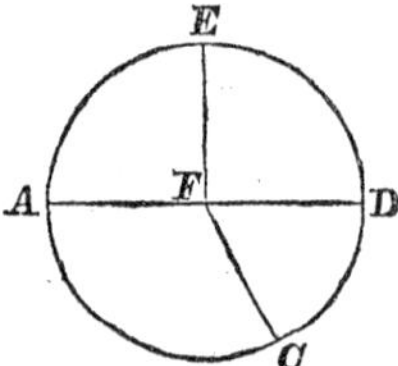


4. *An Arc* is any part of the circumference of a circle. Thus, *GEH* is an arc.

5. *A Chord* is a line joining the two extremities of the arc of a circle. It divides the circle into two unequal parts. Thus, *GH* is a chord.

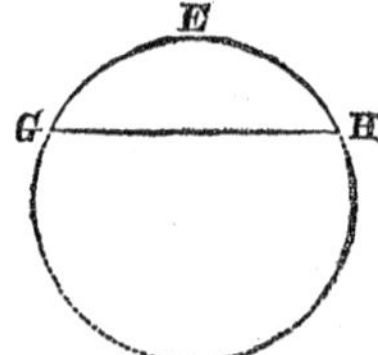


6. *A Segment* is that part of the area of a circle contained between an arc and its chord. It is the part of a circle cut off by a chord. Thus, the space *GHE* is a segment.

7. *A Sector* is a part of a circle comprehended between two radii and the included arc. Thus *AFC* and *CFH* are sectors.

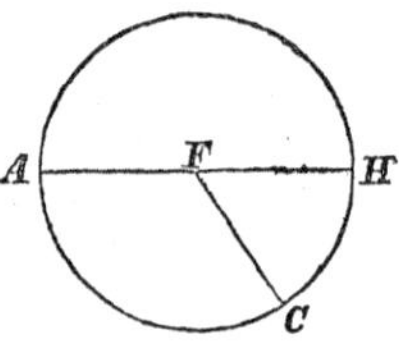


---

¶ **31.** Topic. A circle. Bounding line. Center. Note. Diameter. Semi-circle. Radius. Radii. Their equality. An arc. A chord. Seg-

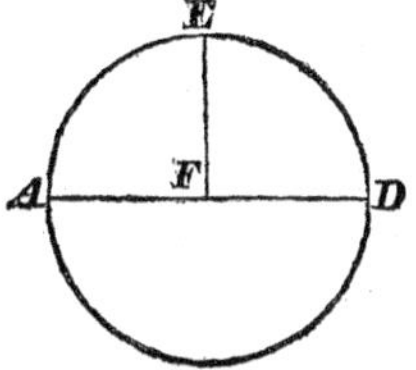


8. *A Quadrant* is the quarter of a circle, or of the circumference of a circle. Thus, *AFE* and *EFD* are quadrants.

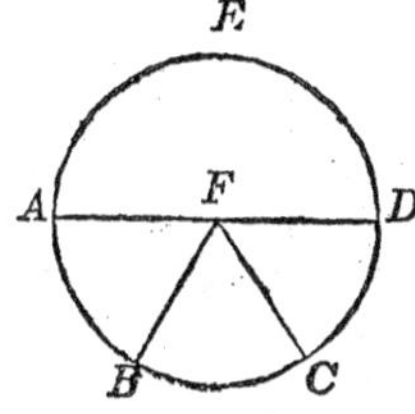


9. *A Sextant* is the sixth part of a circle. Thus, *AFB*, *BFC*, and *CFD*, are sextants.

10. The circumference of every circle is divided into 360 equal parts, called *Degrees;* each degree into 60 equal parts, called *Minutes;* and each minute into 60 equal parts, called *Seconds.*

90°
360° 180°
270°

11. Degrees, minutes, and seconds, are marked respectively °, ′, ″; they are used in mensuration and geometry, for the measurement of angles.

12. Every semi-circle contains 180°, every quadrant 90°, and every sextant 60°.

13. If two lines perpendicular to each other cross in the center of a circle, and terminate in its circumference, they will divide the circle into four equal parts, or quadrants, each having a right angle at the center. Hence, every right angle contains 90 degrees.

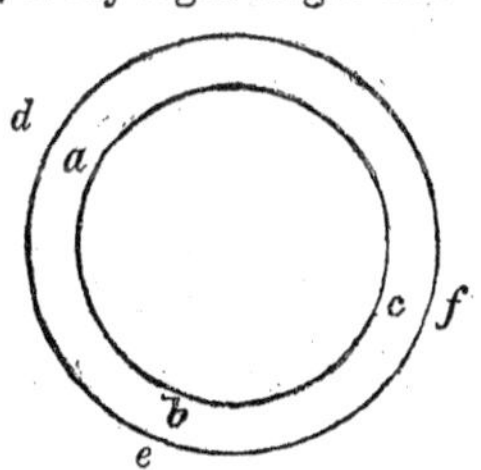


14. *Concentric Circles* are circles of different radii, having a common center. Thus, *a b c* and *d e f* are concentric circles.

---

ment. Sector. Quadrant. Sextant. Divisions of the circumference of circles. Signs. Use of °, ′, and ″. Number of degrees in a circle; — in a

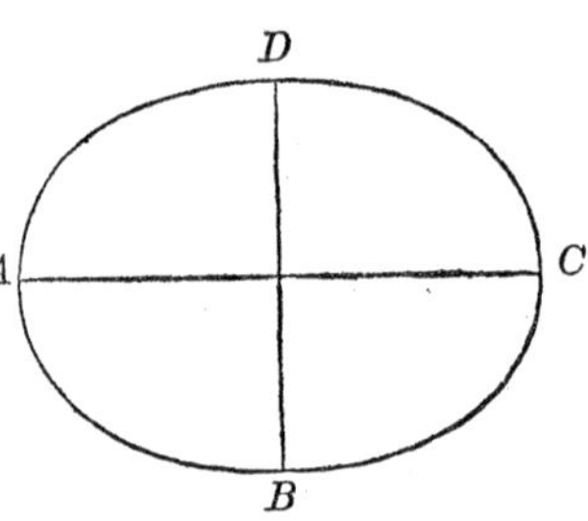


15. *An Ellipse* is an oval figure, bounded by one continuous curve. It has two diameters, the longer of which is called the *Transverse*, and the shorter the *Conjugate* diameter. The two diameters are also called the *Axes*. Thus, *AC* is the transverse, and *BD* the conjugate diameter of the ellipse *ABCD*.

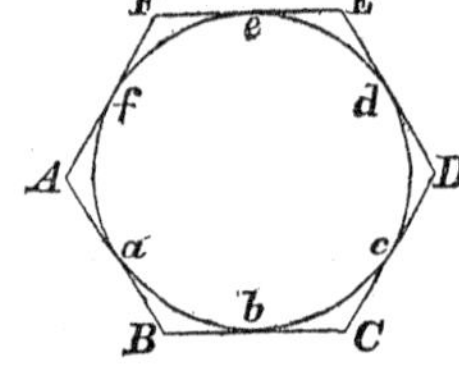


16. A circle, so drawn in a polygon that its periphery touches all the sides of the polygon, is said to be *inscribed in the polygon*, and the polygon is said to be *circumscribed about the circle*. Thus, the circle *a b c d e f* is inscribed in the polygon *ABCDEF*, and the polygon is circumscribed about the circle.

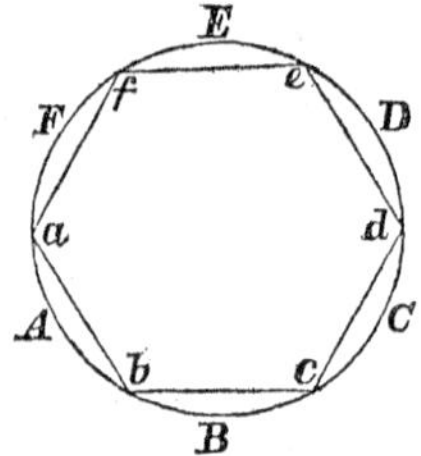


17. A polygon, so drawn in a circle that each of its angles stands on the periphery of the circle, is said to be *inscribed in the circle*, and the circle is said to be *circumscribed about the polygon*. Thus, the polygon *a b c d e f* is inscribed in the circle, and the circle is circumscribed about the polygon.

NOTE 2. Each of the regular polygons may be inscribed in, or circumscribed about, a circle.

---

## SOLIDS OR BODIES.

**¶ 32.** 1. *A Solid* or *a Body* is a magnitude which has length, breadth, and thickness.

2. *A Polyedron* is a solid bounded by many faces or planes.

---

semicircle; — in a quadrant; — in a sextant. Proof that every right angle contains 90°. Concentric circles. An ellipse. Its diameters. Its axes. Circle inscribed in a polygon. Polygon circumscribed about a circle. Polygon inscribed in a circle. Circle circumscribed about a polygon. Note 2.

**¶ 32.** Topic. A solid or a body. Polyedron. Regular solids. Solid

3. *A Regular Solid* is one whose faces are all regular polygons, similar and equal to each other.

4. *A Solid Angle* is one made by the meeting of more than two plane surfaces at one point.

5. *Similar Solids* are such as are contained by the same number of similar planes, similarly situated, and having like angles.

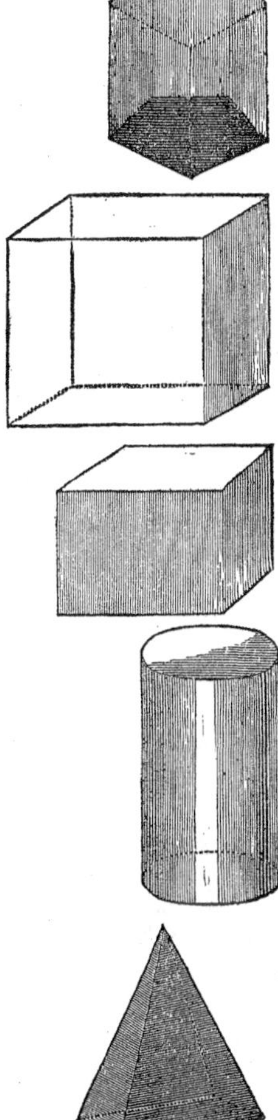

6. *A Prism* is a solid whose bases or ends are any similar, equal, and parallel plane figures, and whose sides are parallelograms.

7. *A Cube* is a solid bounded by six equal squares. The cube is sometimes called the *Right Prism.*

8. *A Parallelopiped* is a solid bounded by six parallelograms, the opposite ones of which are parallel and equal to each other. Or, it is a prism whose base is a parallelogram.

9. *A Cylinder* is a long, circular body, of uniform diameter, its extremities being equal parallel circles.

10. *A Cylindroid* is a solid which differs from the cylinder in having ellipses instead of circles for its ends or bases.

11. *A Pyramid* is a solid whose base is a polygon, and whose sides are triangles terminating in a point called the *Vertex.*

12. *The Segment of a Pyramid* is a part cut off by a plane parallel to the pyramid's base.

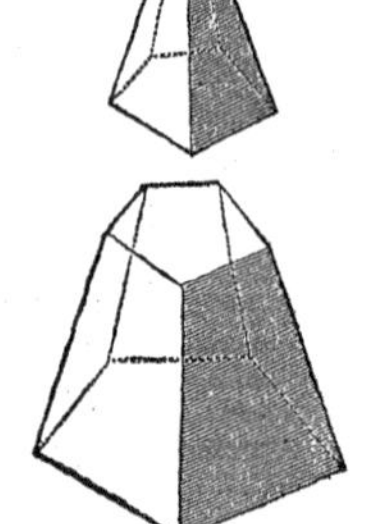

13. *The Frustrum of a Pyramid* is the part left, after cutting off a segment.

14. *A Cone* is a solid whose base is a circle, and whose top terminates in a point or vertex.

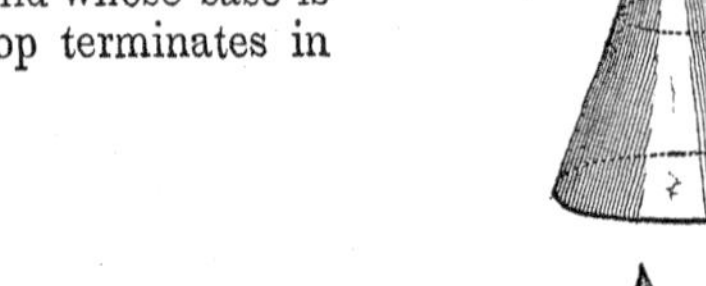

15. *The Segment of a Cone* is a part cut off by a plane parallel to the cone's base.

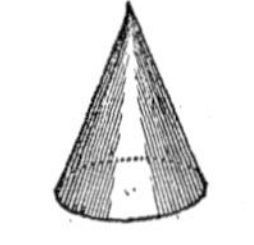

16. *The Frustrum of a Cone* is the part left, after cutting off a segment.

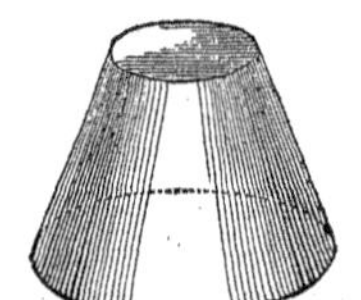

17. *A Sphere or Globe* is a solid bounded by a single surface, which in every part is equally distant from a point called its center.

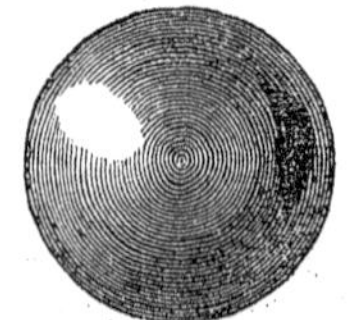

---

angle. Similar solids. Prism. Cube. Parallelopiped. Cylinder. Cylindroid. Pyramid. Its vertex. Segment of a pyramid. Frustrum of a pyramid. Cone. Segment of a cone. Frustrum of a cone. Sphere or globe.

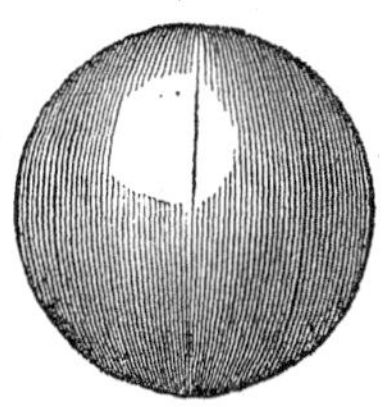

18. *The Axis of a Sphere* is a right line, real or imaginary, passing through its center, on which it does or may revolve.

19. *The Diameter of a Sphere* is a right line passing through its center, and terminating at its surface.

20. *The Radius of a Sphere* is its semi-diameter.

If a sphere be divided into two equal parts, by a plane passing through its center, the parts will be called *Hemispheres.* Hence,

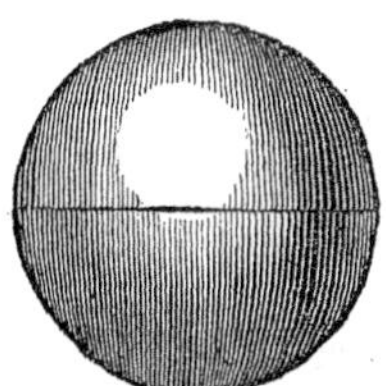

21. *A Hemisphere* is one half of a sphere or globe.

The regular solids are five in number; the *Tetraedron*, the *Hexaedron*, the *Octaedron*, the *Dodecaedron*, and the *Icosaedron.*

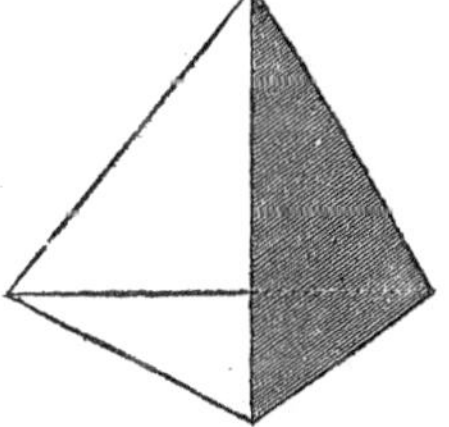

22. *The Tetraedron* is a triangular pyramid, bounded by four equal and equilateral triangles.

23. *The Hexaedron* or *Cube* is a solid bounded by six equal squares.

---

**Its axis. Its diameter. Its radius. Hemisphere. Classification of the reg-**

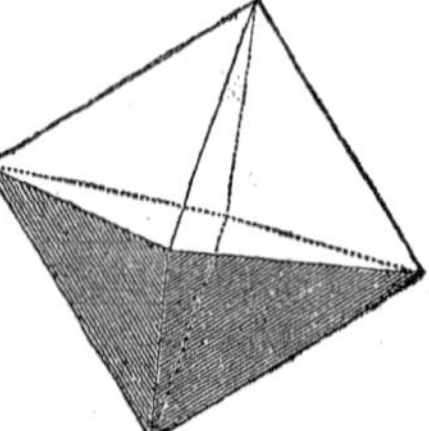

24. *The Octaedron* is a solid bounded by eight equal and equilateral triangles.

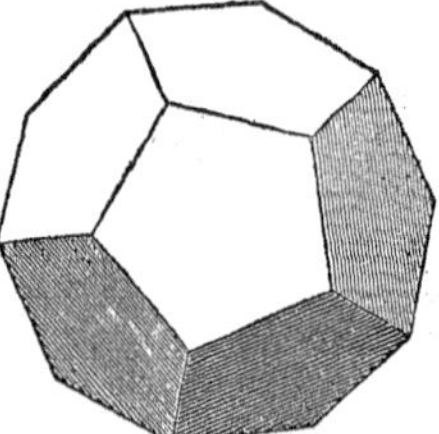

25. *The Dodecaedron* is a solid bounded by twelve equal regular pentagons.

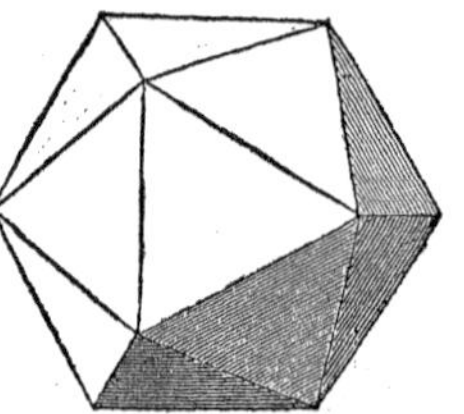

26. *The Icosaedron* is a solid bounded by twenty equal and equilateral triangles.

NOTE 1. Each of the regular solids may be so contained within a sphere that its angles would all stand on the superficies of the sphere.

NOTE 2. All the angles of a regular solid must be equal to each other.

---

ular solids. Tetraedron. Hexaedron. Octaedron. Dodecaedron. Icosaedion. Note 1. Ncte 2.

# PRACTICAL GEOMETRY.

**¶ 33.** *A Problem* is a proposition or a question proposed, which requires some operation to be performed; as, to describe or draw any of the Geometrical figures.

Performing the operation is called *Solving the problem.*

*Practical Geometry* explains the methods of constructing or describing the geometrical figures.

**¶ 34.** Some instruments will be necessary to the successful prosecution of this subject. A case of drafting instruments will best answer the purpose, but when these cannot be obtained, the dividers or compasses, a common ruler, and a scale of equal parts, will be found sufficient for the solution of all the geometrical problems contained in this work.

The dividers are so well known that a description of them is deemed unnecessary.

The ruler may be any convenient length from 12 to 18 inches, from 1 to 2 inches in width, and from $\frac{1}{8}$ to $\frac{3}{8}$ of an inch in thickness.

The scale of equal parts may be conveniently constructed on one side of the common ruler, as follows: Lay off any portion of one side of the ruler, say 10 inches, into 10 equal parts, thus making each part $\frac{1}{10}$ of the length of the scale, or 1 inch in length. Number these parts in their order from left to right; thus, 1, 2, 3, 4, &c. Then lay off one of these parts into 10 other equal parts, each part being $\frac{1}{10}$ of an inch, or $\frac{1}{100}$ of the length of the scale. Number these parts in their order from left to right, and the scale will be completed.

---

## ¶ 35. Geometrical Problems.

PROBLEM I.

*To draw a line through a given point parallel to a given line.*

---

¶ **33.** Topic. A problem. Solving a problem. Practical Geometry.

¶ **34.** Topic. Instruments necessary for the solution of the geometrical problems. The dividers or compasses. The ruler. Construction of the scale of equal parts.

Let $AB$ be the given line, and $C$ the given point. With $C$ as a center, and any convenient radius greater than the shortest distance from $C$ to $AB$, as $CD$, describe an arc $DF$ indefinitely. With the same radius, and $D$ as a center, describe the arc $CG$. Then make $DF = CG$, and draw the line $CF$, which will be parallel to $AB$.

## Problem II.

*To bisect a given line, or to divide it into two equal parts.*

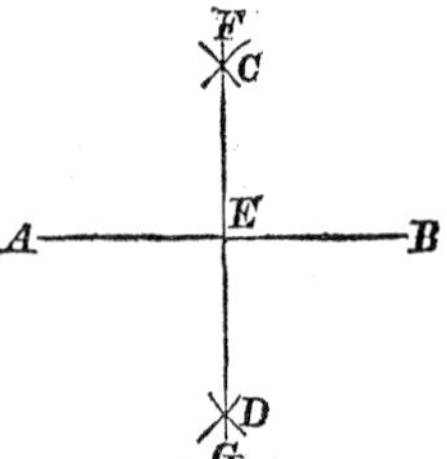


Let $AB$ be the given line. With $A$ as a center, and any radius greater than half of $AB$, describe arcs above and below $AB$, as at $C$ and $D$. With the same radius, and $B$ as a center, describe arcs above and below $AB$, intersecting the arcs first drawn, at $C$ and $D$. Draw the line $FG$ through the points $C$ and $D$, and it will divide the line $AB$ at $E$, into two equal parts $AE$ and $BE$.

## Problem III.

*To bisect a given curve.*

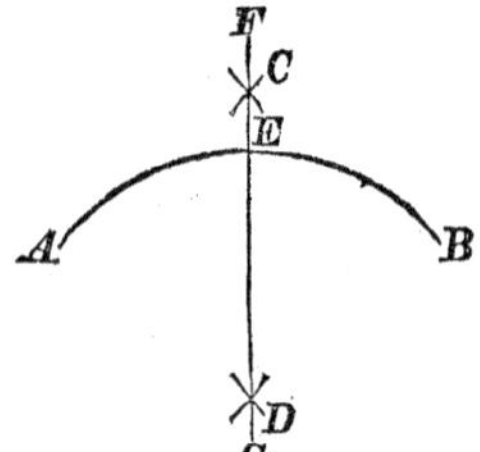


Let $AB$ be the given curve. With $A$ and $B$ as centers, and any radius greater than half of $AB$, describe arcs above and below $AB$, intersecting each other at $C$ and $D$. Draw the line $FG$ through the points $C$ and $D$, and it will bisect the curve $AB$, at $E$.

## Problem IV.

*To bisect a given angle.*

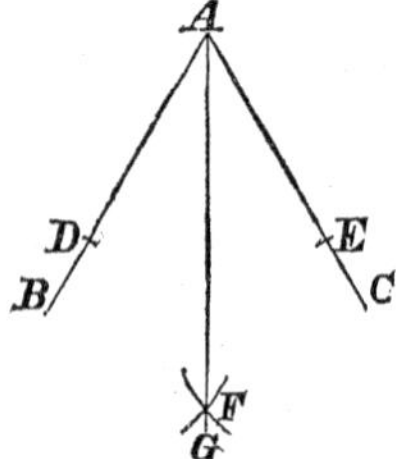


Let $BAC$ be the given angle. Lay off upon $AB$ and $AC$ two points, equally distant from $A$, as $D$ and $E$. With $D$ and $E$ as centers, and any radius greater than half of $DE$, describe two arcs intersecting at $F$. Then draw the line $AG$ through the points $A$ and $F$, and it will bisect the angle $BAC$.

## Problem V.

*To erect a perpendicular on the middle of a given line.*

Let $AB$ be the given line. Bisect the line $AB$, by Prob. II. Then the line $FE$ will be perpendicular to, and will stand on the middle of the line $AB$.

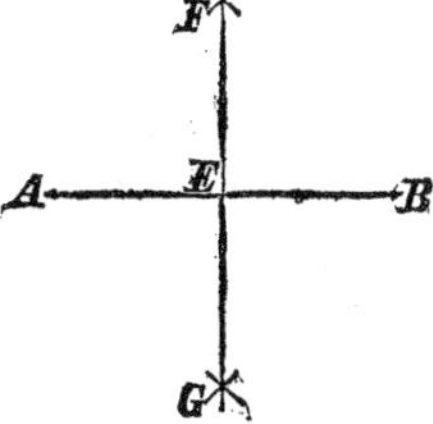


## Problem VI.

*To erect a perpendicular on any given point in a line.*

Let $E$ be the given point, and $AB$ the given line. From $E$ lay off any two equal distances, $EG$ and $EH$, upon the line $AB$. With $G$ and $H$ as centers, and any radius greater than $EG$, describe two arcs intersecting each other in $C$. Then draw the line $FE$, and it will be the required perpendicular.

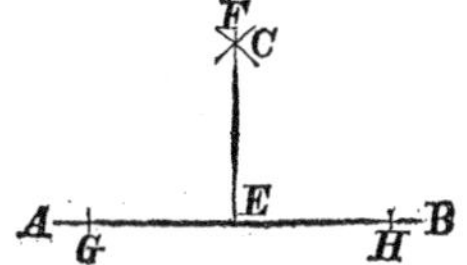


## Second Method.

Let $B$ be the given point, and $AB$ the given line. With any point $C$ as a center, and a radius equal to $BC$, describe the semicircle $DBE$. Draw the diameter $DE$ through the points $D$ and $C$. Then draw a line from $B$ through the point $E$, and it will be the required perpendicular.

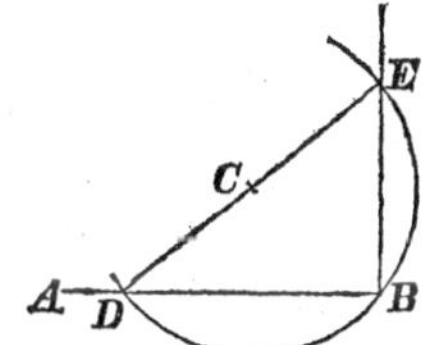


Note. The second method of solving this problem is based upon the principle that all angles in a semicircle are right angles.* In erecting a perpendicular on or near the end of a line, the second method is preferable to the first.

## Problem VII.

*From any point without a given line to draw a perpendicular to the line.*

---

* Euclid's Elements of Geometry.

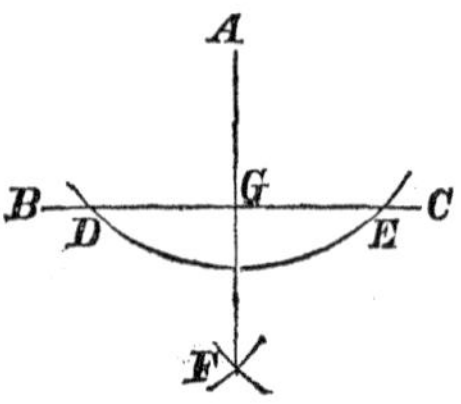


Let $A$ be the given point, and $BC$ the given line. With $A$ as a center, and any radius greater than the shortest distance from $A$ to the line $BC$, describe an arc intersecting $BC$ in two points, $D$ and $E$, which are equi-distant from $A$. With $D$ and $E$ as centers, and the radius $AD$, describe two arcs intersecting each other in $F$. Then draw the line $AF$, and $AG$ will be the required perpendicular.

## Problem VIII.

*To describe a circle which shall pass through any three given points not in a right line.*

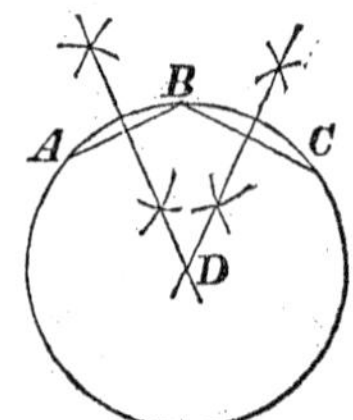


Let $A$, $B$, and $C$ be the given points. Connect the points $A$ and $B$, and the points $B$ and $C$, by the lines $AB$ and $BC$. Bisect the lines $AB$ and $BC$, by Prob. II., and the point $D$, where the bisecting lines cross each other, will be the center of the circle. Then with the radius $DA$, $DB$, or $DC$, describe a circle which will pass through the points $A$, $B$, $C$.

## Problem IX.

*To find the center of a circle.*

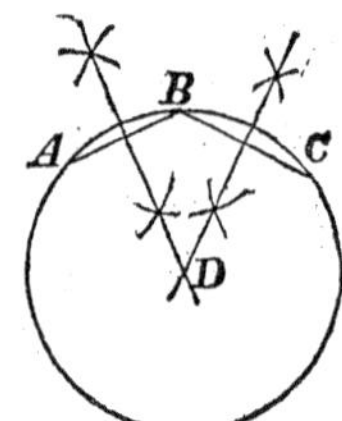


Take any three points in the circumference, as $A$, $B$, $C$, and connect them by the chords $AB$ and $BC$. Bisect the chords $AB$ and $BC$ by Prob. II., and the point $D$, where the bisecting lines cross each other, will be the center of the circle.

## Problem X.

*To find the center of a circle of which an arc only is given.*

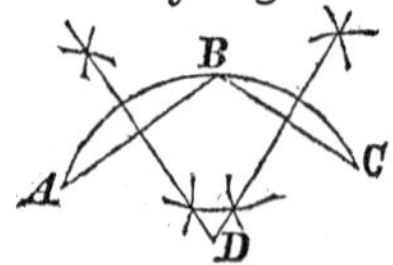


Let $AC$ be the given arc. Take any point in the arc, as $B$, and connect it with the extremities of the arc by the chords $AB$ and $BC$. Bisect these chords by Prob. II., and the point $D$, where the bisecting lines cross each other, will be the center of the circle.

## Problem XI.

*To draw a curve through a given point parallel to a given curve.*

Let *AB* be the given curve or arc, and *C* the given point. First find the center of the circle of which the curve *AB* is an arc, by Prob. X. Then, with *D* as a center, and a radius equal to *DC*, describe the arc *EF*, which will be parallel to the arc *AB*.

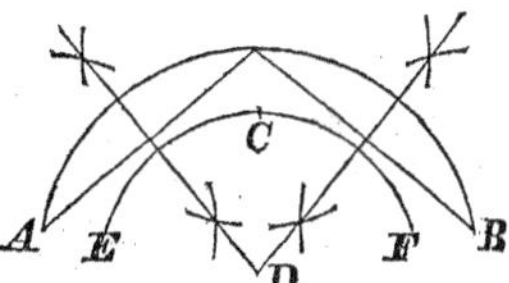


## Problem XII.

*The base and perpendicular of a right-angled triangle being given, to describe the triangle.*

Let *D* be the given base, and *E* the perpendicular. Draw the base *AB* equal to the line *D*. Upon the point *B* erect the perpendicular *BC*, equal to the line *E*, by Prob. VI., and draw the line *AC*. Then the triangle *ABC* will be the required triangle.

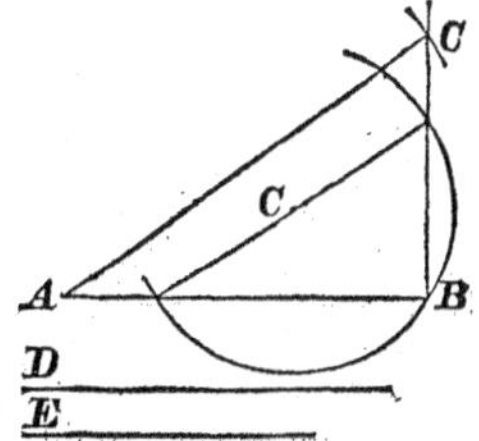


## Problem XIII.

*To describe an equilateral triangle upon a given line or side.*

Let *AB* be the given line or side. With *A* and *B* as centers, and the radius *AB*, describe two arcs intersecting each other in *C*. Then draw the lines *AC* and *BC* and *ABC* will be the required triangle.

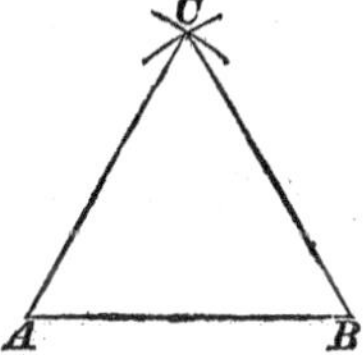


## Problem XIV.

*The three sides of a triangle being given, to describe the triangle.*

Let *A*, *B*, and *C* be the given sides. Draw *DE* equal to the line *A*. With *D* as a center, and a radius equal to the line *B*, and with *E* as a center, and a radius equal to the line *C*, describe arcs intersecting each other in *F*. Draw the lines *DF* and *EF*, and *DEF* will be the required triangle.

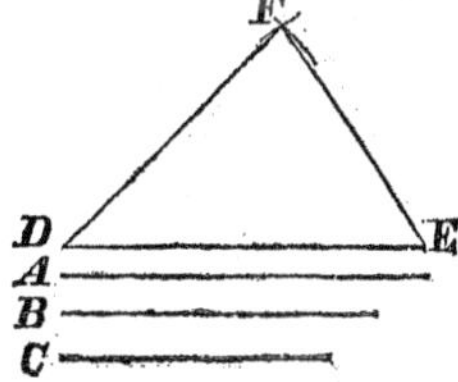

## Problem XV.

*The hypotenuse and one side of a right-angled triangle being given, to describe the triangle.*

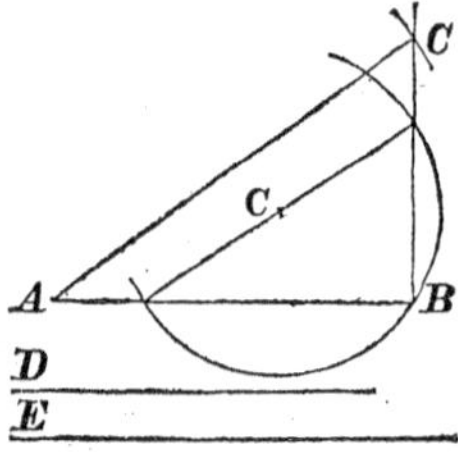


Let *D* be the given side, and *E* the given hypotenuse. Draw the side *AB* equal to the line *D*, and upon the point *B* erect the perpendicular *BC* indefinitely. With *A* as a center, and a radius equal to the line *E*, describe an arc intersecting the perpendicular *BC*, at *C*. Then draw the hypotenuse *AC*, and *ABC* will be the required triangle.

## Problem XVI.

*At a given point to make an angle equal to a given angle.*

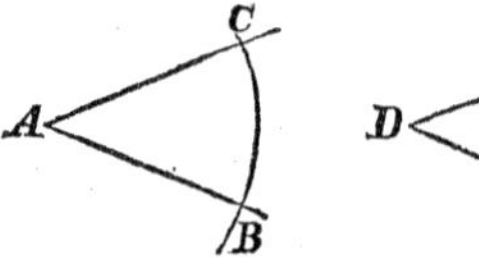


Let *D* be the given point, and *BAC* the given angle. Draw the line *DE* indefinitely. With *A* as a center, and any convenient radius, draw the arc *BC* terminating in the sides of the angle. With the same radius, and *D* as a center, draw the arc *EF*. With *E* as a center, and a radius equal to *BC*, draw an arc intersecting the arc *EF* at *F*. Then through the points *D* and *F* draw the line *DF*, and the angle *EDF* will be equal to the angle *BAC*.

## Problem XVII.

*Two sides of a triangle and the angle which they contain being given, to describe the triangle.*

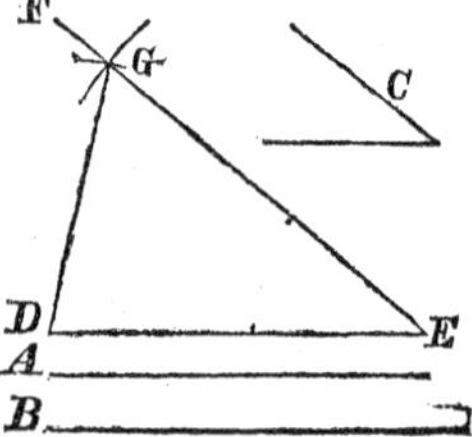


Let *A* and *B* be the given sides, and *C* the given angle. Draw the side *DE* equal to the line *A*. At the point *E* make an angle equal to the angle *C*, by Prob. XVI., and draw the line *EF* indefinitely. With *E* as a center, and a radius equal to the line *B*, describe an arc intersecting the line *EF* in *G*. Then draw the line *DG*, and *DEG* will be the required triangle.

## Problem XVIII.

*To describe a square upon a given line.*

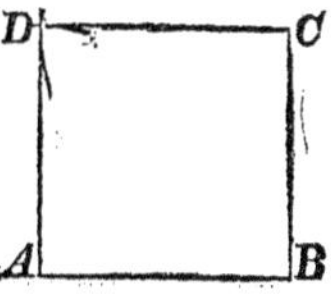


Let *AB* be the given line. At the point *B* erect the perpendicular *BC*, and make it equal to *AB*. With *A* and *C* as centers, and a radius equal to *AB*, describe two arcs intersecting each other in *D*. Then draw the lines *AD* and *CD*, and *ABCD* will be the required square.

## Problem XIX.

*Two adjacent sides of a rectangle being given, to describe the rectangle.*

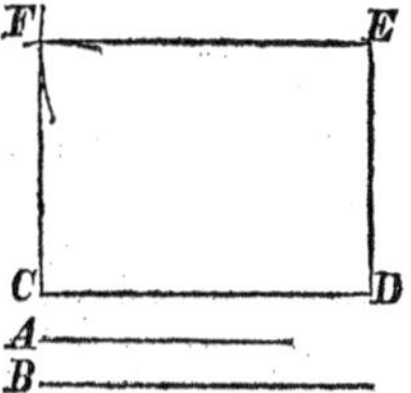


Let *A* and *B* be the given adjacent sides. Draw the side *CD* equal to the line *B*, and upon the point *D* erect the perpendicular *DE* equal to the line *A*. With *C* as a center, and a radius equal to *DE*, describe an arc; and with *E* as a center, and a radius equal to *CD*, describe another arc, intersecting the first at *F*. Then draw the lines *CF* and *EF*, and *CDEF* will be the required rectangle.

## Problem XX.

*One side and one of the angles of a rhombus being given, to describe the rhombus.*

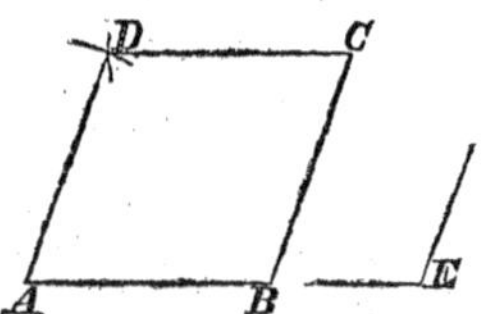


Let *AB* be the given side, and *E* the given angle. At the point *B* make an angle equal to the angle *E*, by Prob. XVI., and draw the line *BC* equal to *AB*. With *A* and *C* as centers, and a radius equal to *AB*, describe two arcs intersecting each other in *D*. Then draw the lines *AD* and *CD*, and *ABCD* will be the required rhombus.

Note. A rhomboid may be readily described, by combining Problems XIX. and XX.

## Problem XXI.

*To inscribe an equilateral triangle in a given circle.*

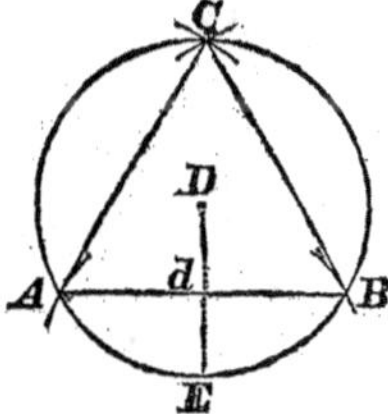


With any point in the circumference, as *E*, for a center, and the radius *DE*, describe two arcs intersecting the circle in *A* and *B*. With *A* and *B* as centers, and a radius equal to *AB*, describe two arcs intersecting each other in *C*. These arcs will intersect each other and the circle in the same point. Then draw the lines *AB*, *BC*, and *CA*, and *ABC* will be the required triangle.

## Problem XXII.

*To inscribe a square in a given circle.*

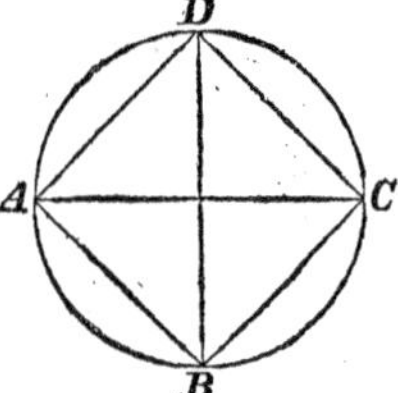


Draw the diameters $AC$ and $BD$ at right angles to each other. Then through the points $A$, $B$, $C$, and $D$, draw the lines $AB$, $BC$, $CD$, and $DA$, and $ABCD$ will be the required square.

## Problem XXIII.

*To inscribe a pentagon in a given circle.*

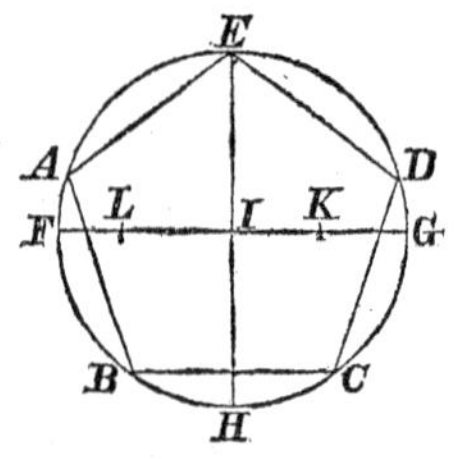


Draw the diameters $FG$ and $EH$, at right angles to each other, and bisect the radius $IG$ at $K$. With $K$ as a center, and a radius equal to $EK$, describe an arc intersecting $FG$ in $L$. Apply the distance $EL$ around the circle, and it will divide it into five equal parts. Then draw the lines $AB$, $BC$, $CD$, $DE$, and $EA$, and $ABCDE$ will be the required pentagon.

## Problem XXIV.

*To inscribe a hexagon in a given circle.*

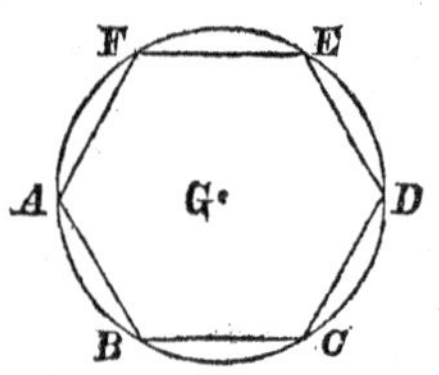


Apply the radius $AG$ around the circle, and it will divide it into six equal parts. Then draw the lines $AB$, $BC$, $CD$, $DE$, $EF$, and $FA$, and $ABCDEF$ will be the required hexagon.

## Problem XXV.

*To inscribe an octagon in a given circle.*

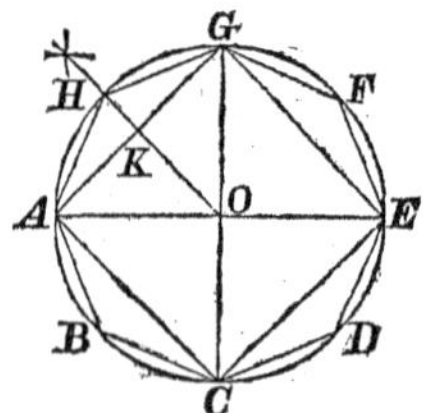


Inscribe the square $ACEG$, by Prob. XXII., and bisect the arcs $AC$, $CE$, $EG$, and $GA$, at $B$, $D$, $F$, and $H$, respectively. Then draw the lines $AB$, $BC$, $CD$, $DE$, $EF$, $FG$, $GH$, and $HA$, and $ABCDEFGH$ will be the required octagon.

## Problem XXVI.

*To inscribe a decagon in a given circle.*

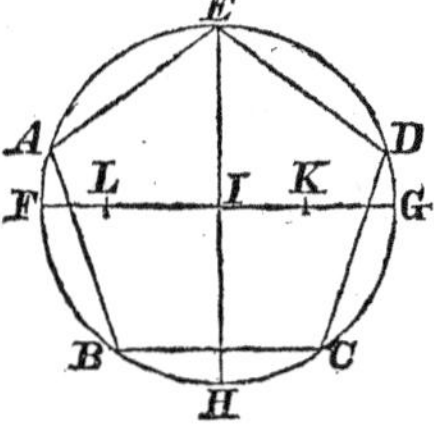


Inscribe the pentagon *ABCDE*, by Prob. XXIII., and bisect the arcs *AB*, *BC*, *CD*, *DE*, and *EA*. Then draw lines through the angles of the pentagon and the points of bisection, and the figure will be the required decagon.

## Problem XXVII.

*To inscribe a dodecagon in a given circle.*

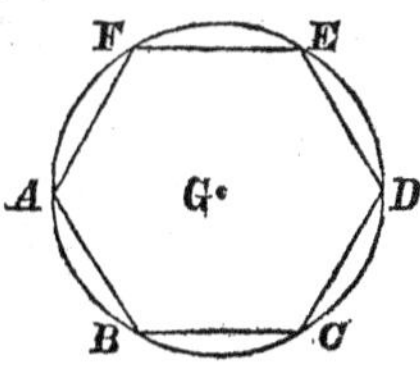


Inscribe the hexagon *ABCDEF*, by Prob. XXIV., and bisect the arcs *AB*, *BC*, *CD*, *DE*, *EF*, and *FA*. Then draw lines through the angles of the hexagon and the bisecting points, and the figure will be the required dodecagon.

## Problem XXVIII.

*To inscribe any regular polygon in a given circle.*

Divide the circle into as many equal parts as the required polygon is to contain sides, and draw lines through the points of division. The inscribed figure will be the required polygon.

## Problem XXIX.

*One side and the number of sides of a regular polygon being given, to describe the polygon.*

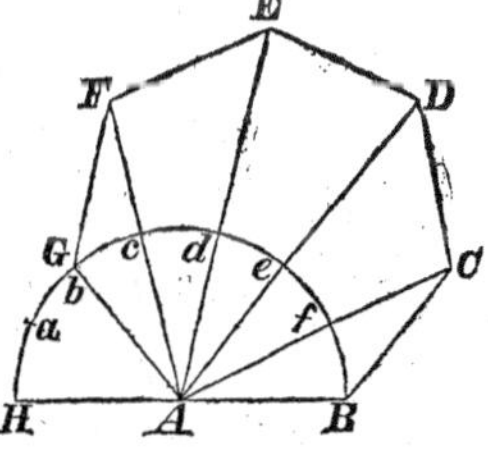


Let it be required to describe a heptagon upon the line *AB*. With the center *A*, and the radius *AB*, describe the semicircle *HabcdefB*, and divide it into seven equal parts. To the second point of division *b*, draw the line *AG*, and through the points *c*, *d*, *e*, and *f*, draw the lines *AF*, *AE*, *AD*, and *AC*. Apply the distance *AB*, from *B* to *C*, from *C* to *D*, from *D* to *E*, from *E* to *F*, and from *F* to *G*. Then draw the lines *BC*, *CD*, *DE*, *EF*, and *FG*, and *ABCDEFG* will be the required heptagon.

Proceed in the same manner with any other regular polygon.

## Problem XXX.

*To circumscribe a regular polygon about a given circle.*

Let it be required to circumscribe a hexagon about a circle. In the given circle inscribe the hexagon *ABCDEF.* To the length of the radius *OA*, add the distance *Pp*, and with this radius, and *O* as a center, describe a second circle. Then in this circle describe the hexagon *abcdef*, and it will circumscribe the given circle.

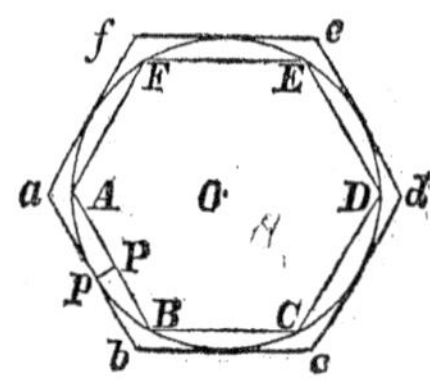


Any other regular polygon may be circumscribed about a circle in the same manner.

## Problem XXXI.

*To circumscribe a circle about a regular polygon.*

Bisect any two adjacent sides of the polygon, as *AB* and *BC*, and the point *D*, where the bisecting lines cross each other, will be the center of the circle. Then, with *D* as a center, and a radius equal to the distance from *D* to any angle of the polygon, as *A*, describe a circle, and it will circumscribe the given polygon.

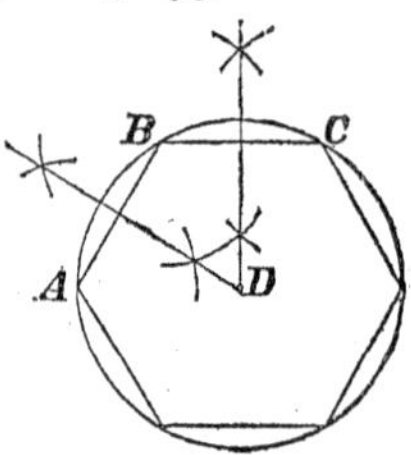


## Problem XXXII.

*To inscribe a circle in a regular polygon.*

Bisect any two adjacent sides of the polygon, as *AB* and *BC*, and the point *D*, where the bisecting lines cross each other, will be the center of the circle. Then, with *D* as a center, and the radius *DE*, describe a circle, and it will be inscribed in the given polygon.

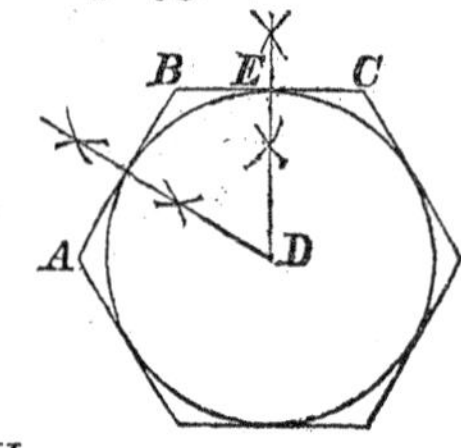


## Problem XXXIII.

*To inscribe a circle in a given triangle.*

Let *ABC* be the given triangle. Bisect any two angles, as *A* and *B*, and the point *D*, where the bisecting lines cross each other, will be the center of the circle. From this point let fall a perpendicular upon one of the sides, as *DE*. Then, with the center *D*, and the radius *DE*, describe a circle, and it will be inscribed in the given triangle.

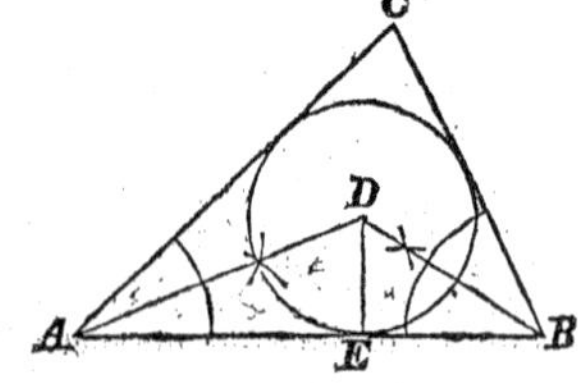

## Problem XXXIV.

*To construct solids.*

Upon pasteboard, or any other pliable matter, draw figures like the following. Cut the bounding lines entirely through, and the other lines half through; turn up the sides and glue the edges together, and the figures will form the solids named below.

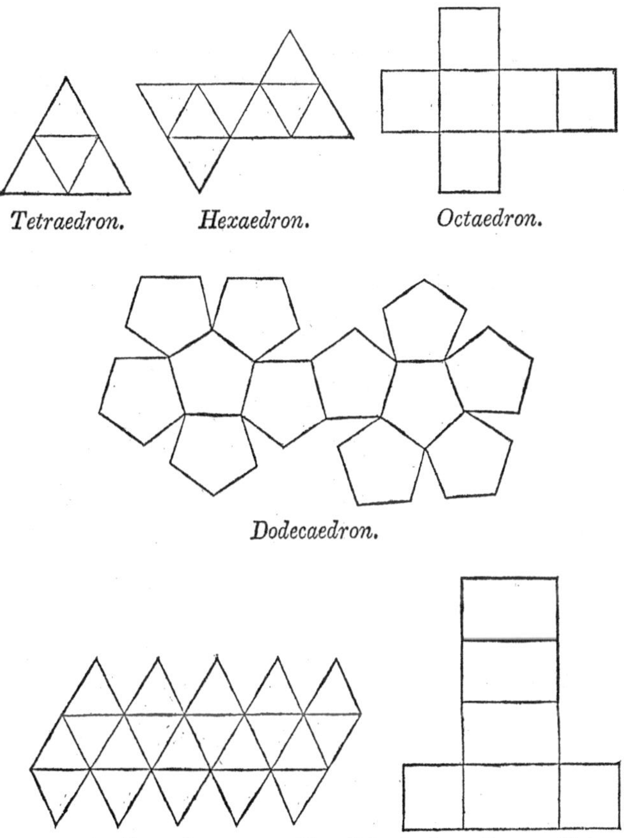

*Tetraedron.* *Hexaedron.* *Octaedron.*

*Dodecaedron.*

*Icosaedron.* *Parallelopiped, or Square Prism.*

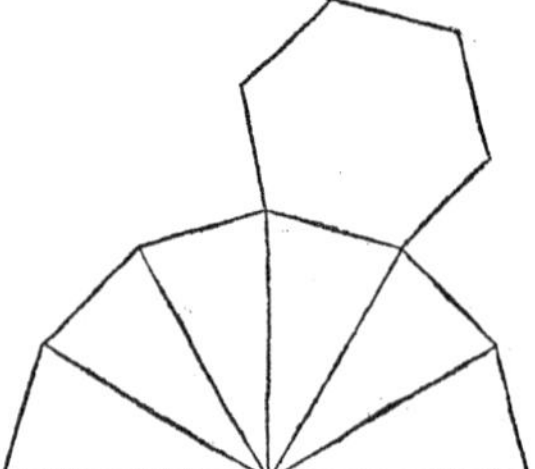

*Hexagonal Pyramid.*

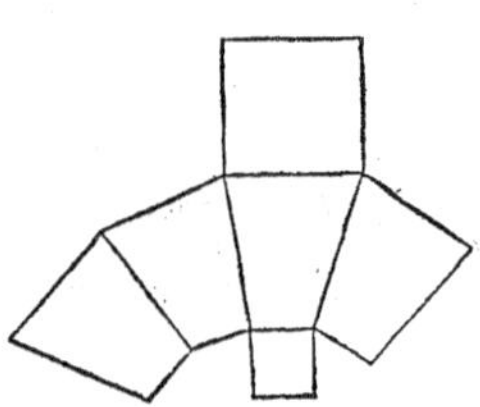

*Frustrum of a Square Pyramid.*

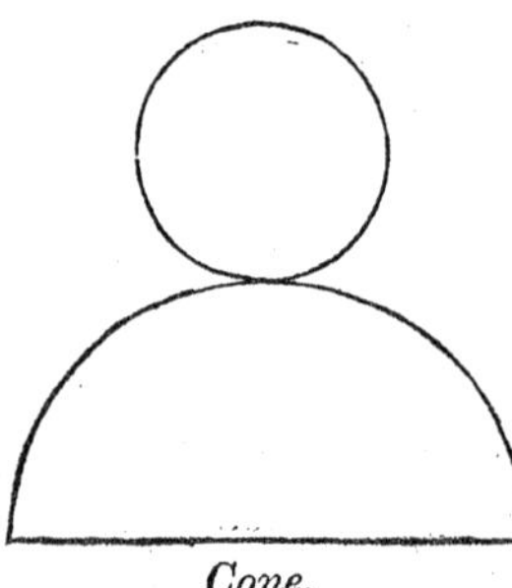

*Cone.*

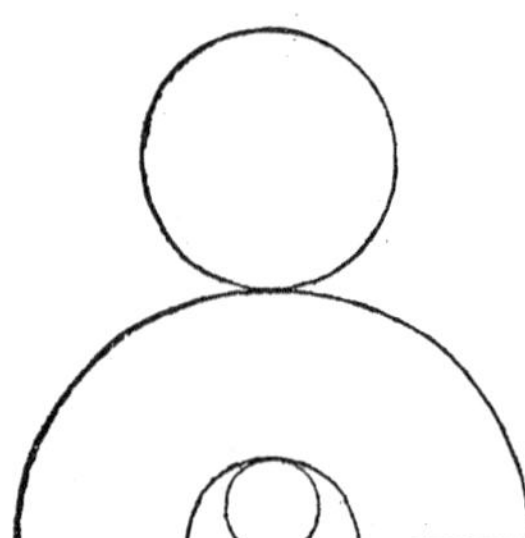

*Frustrum of a Cone.*

## MENSURATION OF LINES AND SUPERFICIES.

**¶ 36.** *The area of a figure* is its superficial contents, or the surface included within any given lines, without regard to thickness.

In taking the dimensions of any line, surface, or solid, we are always governed by some denomination, a unit of which is called the *Unit of Measure.* Thus, if any lineal measure be estimated in feet, the unit of measure is 1 foot; if in inches, the unit is 1 inch; if in yards, the unit is 1 yard, &c. If any superficial measure be estimated in feet, the unit of measure is 1 square foot, or 144 square inches; if in yards, the unit is 1 square yard, or 9 square feet, &c. If any solid or cubic measure be estimated in feet, the unit of measure is 1 cubic foot, or 1728 cubic inches; if in yards, the unit is 1 cubic yard, or 27 cubic feet, &c.

**¶ 37.** *The length and breadth of a square or rectangle being given, to find the square contents.*

### RULE.

Multiply the length by the breadth, and the product will be the square contents.

NOTE. For an analysis of the principles of this and the following rule, see Revised Arithmetic, ¶¶ 48, 49, and 50.

### EXAMPLES FOR PRACTICE.

1. How many square inches in a board 16 inches square?

2. How many square rods in a field 90 rods long, and 52 rods wide? How many acres? *Ans. to last,* 29 A. 1 R.

3. A certain village lot of land is 66 feet front, by 330 feet deep; how many poles does it contain?

4. In a field 220 rods long and 90 rods wide, how many acres?

5. A certain rectangular piece of land measures 1000 links by 100; how many chains does it contain? How many acres? *Ans.* 1 acre.

**¶ 38.** *The square contents or area, and one side of a square or rectangle being given, to find the other side.*

---

**¶ 36.** Topic. The area of a figure. Unit of measure. Examples.
**¶ 37.** Topic. Analysis. Rule.

**RULE.**

Divide the square contents by the given side, and the quotient will be the required side.

NOTE. The area and the given side must be reduced to corresponding denominations before dividing; that is, if the area is expressed in square feet, the given side must be in feet. *Or*, they may be reduced to any other corresponding denominations, as inches and square inches, yards and square yards, rods and square rods, &c.

**EXAMPLES FOR PRACTICE.**

1. If a piece of land 20 rods in length contain 240 square rods, what is its width?
2. The side of a certain building 16 feet in hight contains 2560 square feet; what is its length?
3. What length of carpeting 5 quarters wide is equal to a square yard? *Ans.* 3'2 qrs. = 28'8 in.
4. How many yards of cloth $1\frac{3}{8}$ yards wide are equal to 15 yards $\frac{3}{4}$ of a yard wide? *Ans.* $8\frac{2}{11}$ yds.
5. A piece of land 8 chains wide contains 40 acres; what is its length in chains? *Ans.* 50 chains.

**¶ 39.** *The base and perpendicular of a right-angled triangle being given, to find the hypotenuse.*

**RULE.**

Square the base and the perpendicular, add the squares together, and extract the square root of their sum; the root will be the length of the hypotenuse.

NOTE 1. For an analysis of the principles upon which this and the following rule are founded, see Revised Arithmetic, ¶ 210, Note 3.

**EXAMPLES FOR PRACTICE.**

1. The base is 12 inches, and the perpendicular 5 inches; what is the hypotenuse?
2. The gable of a house is 28 feet wide, and the perpendicular hight of the ridge of the roof above the eaves is 7 feet; what is the length of the rafters?

NOTE 2. The gable is the portion above a horizontal line extending from one eave to the other. Thus, *ABC* may represent the gable of a house, and may readily be divided into two right-angled triangles, *ADB* and *CDB*. *Ans.* 15'65+ feet.

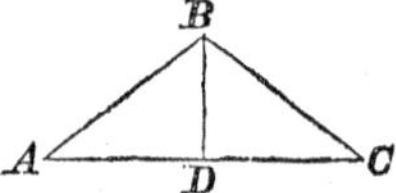


3. Upon a plane 25 feet long stands a pole 12 feet high, at the distance of 9 feet from one end of the plane; what is the length of a rope that will extend from one end of the plane to the other, over the top of the pole? *Ans.* 35 feet.

---

**¶ 38.** Topic. Analysis. Rule. Note.
**¶ 39.** Topic. Analysis. Rule. Note 1. Note 2. Note 3. Note 4.

4. The second floor of a certain house is 9 feet above the first, and each of the steps in the flight of stairs leading from the first floor to the second, is 9 inches high and 12 inches wide; what is the slant hight of the flight of stairs?

NOTE 3. The pupil will perceive that the stairs rise 3-4 as fast as they advance; that is, the perpendicular is 3-4 as long as the base.
NOTE 4. The hypotenuse of any right-angled triangle whose base and perpendicular are to each other as 4 to 3, is equal to the longer side plus ¼ of itself.

5. The base is 20, and the perpendicular 15; what is the hypotenuse?
6. The base is 48, and the perpendicular 64; what is the hypotenuse?
7. The hights of two trees, 75 feet apart, are 96 and 130 feet; how far from the top of one tree to the top of the other? How far from the top of each to the bottom of the other?

*Ans.* { From top of one to top of the other, 82'34+ ft.
" " " taller to bottom of shorter, 150'08 ft.
" " " shorter to bottom of taller, 121'82+ ft. }

**¶ 40.** *The hypotenuse and one leg of a right-angled triangle being given, to find the other leg.*

### RULE.

Square the hypotenuse and the given leg, subtract the square of the leg from the square of the hypotenuse, and extract the square root of the remainder; the root will be the length of the other leg.

NOTE. The pupil will perceive that this rule is the reverse of the one given in the last ¶.

### EXAMPLES FOR PRACTICE.

1. The hypotenuse of a right-angled triangle is 5 feet long, and the base 4 feet; what is the length of the perpendicular?
2. A ladder 17 feet long is so placed that it touches the wall 15 feet above the plane on which the wall and ladder stand; how far from the foot of the wall to the foot of the ladder? *Ans.* 8 feet.
3. The length of the rafters to a certain building is 13 feet, and the perpendicular hight of the ridge above the eaves is 5 feet; what is the width of the gable? *Ans.* 24 feet.
4. One side of the roof of a certain house is 18 feet wide, the other side 16, and the perpendicular hight of the ridge above the eaves 9 feet; what is the width of the gable? *Ans.* 28'8+ feet.
5. A ladder 25 feet long is so placed between two buildings, that when its top is leaned against one of them, it touches the building 20 feet from the ground, and when leaned against the other, it touches it 15 feet from the ground; what is the horizontal distance between the buildings? *Ans.* 35 feet.
6. The distance from the spot on which I stand to the top of a certain tree is 100 feet, and to the bottom of the same, (which is in the same plane with my feet,) 60 feet; how high is the tree? *Ans.* 80 feet.

---

¶ **40.** Topic. Analysis. Rule. Note.

7. The distance from the top of one tree to the top of another standing 75 feet from the first, is 100 feet, and the shorter tree is 80 feet high; what is the hight of the taller tree? *Ans.* 146·14 feet.

**¶ 41.** *The sum and difference of two numbers being given, to find the numbers.*

NOTE. A knowledge of the principles contained in this and the following ¶ is necessary to a clear comprehension of the rule in ¶ 43.

Ex. The sum of two numbers is 25, and their difference is 7; what are the numbers?

ANALYSIS. The sum of two numbers plus their difference will give twice the greater number; and the sum minus the difference will give twice the less number. $25 + 7 = 32$, and $32 \div 2 = 16$, the greater number. $25 - 7 = 18$, and $18 \div 2 = 9$, the less number. Hence,

*The sum and difference of two numbers being given, to find the numbers.*

### RULE.

I. *For the greater number;* — Add the sum and difference together, and divide the amount by 2.

II. *For the less number;* — Subtract the difference from the sum, and divide the remainder by 2.

### EXAMPLES FOR PRACTICE.

1. The sum of two numbers is 92, and their difference is 56; what are the numbers?

2. The sum of the lengths of two lines is 126 yards, and their difference is 31 yards; what is the length of each line?

*Ans.* { Longer line, 78·5 yds.<br>Shorter " 47·5 "

3. Two men own 350 acres of land, and one owns 62 acres more than the other; how many acres does each man own?

*Ans.* { One man owns 206 acres.<br>The other owns 144 "

**¶ 42.** *The sum of two numbers and the difference of their squares being given, to find the numbers.*

Ex. 1. The sum of the numbers 16 and 9 is 25, and their difference is 7; what is the difference of their squares?

ANALYSIS. $16^2 = 256$, $9^2 = 81$, and $256 - 81 = 175$. But 25, the sum of the two given numbers, multiplied by 7, their difference, gives the same result; thus, $25 \times 7 = 175$. Therefore,

---

¶ **41.** Topic. Note. Solution of Ex. 1. Rule.

¶ **42.** Topic. Solution of Ex. 1. Conclusion. Solution of Ex. 2. Rule.

*The product of the sum and difference of two numbers is equal to the difference of their squares.*

Ex. 2. The sum of two numbers is 25, and the difference of their squares is 175; what are the numbers?

Analysis. In this example we have the sum of two numbers, and the difference of their squares given, to find the numbers. That is, we have the product, 175, and one factor, 25, to find the other factor. $175 \div 25 = 7$, the other factor, or the difference of the two numbers. We now have 25, the sum of two numbers, and 7, their difference, to find the numbers, (¶ 41.) $25 + 7 = 32$, and $32 \div 2 = 16$, the greater number. $25 - 7 = 18$, and $18 \div 2 = 9$, the less number. Hence,

*When the sum of two numbers, and the difference of their squares are given, to find the numbers.*

**RULE.**

I. Divide the difference of the squares by the sum of the numbers, and the quotient will be the difference of the numbers.

II. From the sum and difference, find the numbers by ¶ 41, Rule.

**EXAMPLES FOR PRACTICE.**

1. The sum of two numbers is 30, and the difference of their squares is 300; what are the numbers?

2. The sum of two numbers is 56, and the difference of their squares is 112; what are the numbers?

3. The sum of two numbers is 88, and the difference of their squares is 6688; what are the numbers? *Ans.* 82 and 6.

**¶ 43.** *One side of a right-angled triangle, and the sum of the hypotenuse and the other side being given, to find the hypotenuse and the other side.*

Ex. The sum of the hypotenuse and perpendicular of a right-angled triangle is 24 inches, and the base is 12 inches; what is the length of the hypotenuse, and also of the perpendicular?

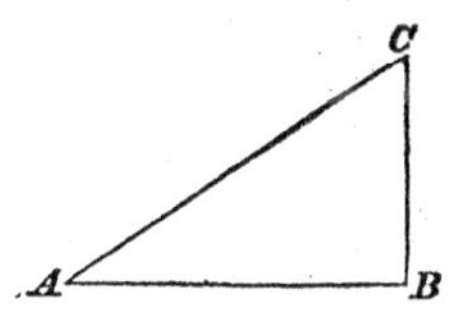


Analysis. The 24 inches is the length of the side $AC$ plus the length of the side $BC$, and consequently is the sum of two numbers. The 12 inches is the length of the side $AB$, and is the square root of the difference between the squares $AC$ and $BC$. We therefore have the sum of two numbers, and the square root of the difference of their squares, to find the numbers. $12^2 = 144$,

¶ **43.** Topic. Solution of Ex. Rule. Note.

the difference of the squares of the two numbers, and 144 ÷ 24 = 6, the difference of the numbers, (¶ **42.**) We now have 24, the sum of two numbers, and 6, their difference, to find the numbers. (¶ **41.**) 24 + 6 = 30, and 30 ÷ 2 = 15, the greater number. 24 − 6 = 18, and 18 ÷ 2 = 9, the less number.

*Ans.* { Length of hypotenuse, 15 inches.
" " perpendicular, 9 "

Hence, *When one leg of a right-angled triangle, and the sum of the hypotenuse and the other leg are given, to find the hypotenuse and the other leg.*

**RULE.**

I. Divide the square of the given side by the sum of the other two sides, and the quotient will be the difference of the two unknown sides.

II. From the sum and difference of the two unknown sides, find the two sides by ¶ 41, Rule.

NOTE. The operation may be proved by ¶ 39, Rule.

**EXAMPLES FOR PRACTICE.**

1. The sum of the lengths of the hypotenuse and base of a right-angled triangle is 25 feet, and the perpendicular is 5 feet; the lengths of the hypotenuse and base are required. *Ans.* { Hypotenuse, 13 ft.
Base, 12 "

2. A gentleman has a triangular park, one of the shorter sides of which measures 24 rods, and the other two sides measure 72 rods; what is the length of each of the other two sides?

3. Two men, A and B, started from a certain place, and traveled, A going east 35 miles, and B north a certain distance, and then in a right line to the place at which B had stopped; the distance that B traveled was $\frac{7}{12}$ of the whole distance that A and B had both traveled; how far did B travel? How far was he from A, and also from the place from which he started, when he stopped his northerly course and turned to meet B? *Answers,* in order. 49 mi.; 37 mi.; 12 mi.

4. A tree 100 feet high was broken off by the wind, at such a distance from the bottom, that the top part touched the ground 50 feet from the foot of the tree, the bottom of the part broken off resting upon the top of the stump; what were the hight of the stump, and the length of the part broken off? *Ans.* { Hight of the stump, 37·5 ft.
Length of part broken off, 62·5 "

**¶ 44.** *The relation of the three sides of a right-angled triangle to each other, applied to the measurement of distances.*

---

¶ **44.** Topic. Solution of Ex. 1.—of Ex. 2. Note. Principles upon which the operations are performed. Solution of Ex. 5.—of Ex. 6.

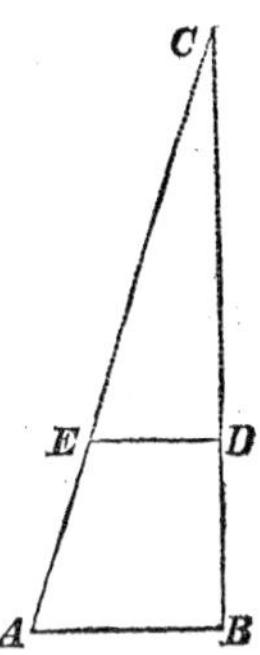


1. In the accompanying figure, the base *AB* is 9 feet long, the perpendicular *BD*, 9 feet, and the line *DE*, 6 feet; what was the whole length of the perpendicular *BC*, of the original triangle *ABC*.

Analysis. It is evident that the line *AE*, in the perpendicular distance of 9 feet, inclines to or approaches the perpendicular *BC* 3 feet. The base of the triangle is 9 feet; and, in order to complete the triangle, the line *AE* must be continued till it meets the perpendicular, that is, till it has approached it 9 feet. Since it approaches it 3 feet in 9, it must be continued 3 times as far to approach it 9 feet, that is, to meet it. $3 \times 9$ feet $= 27$ feet. Therefore, the perpendicular *BC* was 27 feet long.

2. It is required to find the distance to an inaccessible point, *O*, on the opposite side of a river.

Operation. Place a stake at *B*, and retreat to the point *A*, any convenient distance, say 10 feet, so that the points *A*, *B*, and *O*, are in the same right line *AO*. Then, upon the base *AB*, construct the square *ABCD*. Next, from the point *D* take an observation; note on the line *BC* the point at which the line *DO* will intersect the line *BC*, as at *E*; and measure the distance *CE*, say 2 feet. You now have the base *AD*, 10 feet; the perpendicular *AB*, 10 feet; and the distance *CE*, 2 feet; to find the length of the line *AO*. Since the line *DE* approaches the line *AO* 2 feet in 10, in 1 foot it approaches it $\frac{1}{10}$ of 2 feet, or $\frac{2}{10} = \frac{1}{5}$ of a foot.

$\frac{1}{5}$ ft. : 1 ft. : : 10 ft. : 50 ft.; *or*,
2 ft. : 10 ft. : : 10 ft. : 50 ft.

That is, the distance the line *DO* approaches the line *AO* in any given number of feet, is to that number of feet, as the whole line or base *AD* is to the whole length of the line *AO*. 50 feet, the distance from *A* to *O*, minus 10 feet, the distance from *A* to *B*, leaves 40 feet, the distance from *B* to *O*. *Ans.* 40 feet.

Note. The distance to any visible point on a plain, precipice, mountain, tower, &c., may be ascertained, by varying the foregoing principles to suit the circumstances under which the observations are taken. The observations must always be taken upon a surface in the same plane with the object, unless the elevation or depression of the object, above or below the plane on which the observation is taken, be known.

3. Given the perpendicular 10 feet, the base 10 feet, and the inclination of the second line of observation $2\frac{1}{2}$ inches; that is, $2\frac{1}{2}$ inches to 10 feet, the length of the perpendicular; what is the distance from the nearest accessible point to the object? *Ans.* 470 ft.

4. Wishing to ascertain the distance to a tree on the top of a certain hill, without applying the chain, I constructed a square of 10 feet, as in the last two examples, and found the inclination to be $\frac{1}{8}$ of an inch; what was the distance to the tree? *Ans.* 1920 ft.

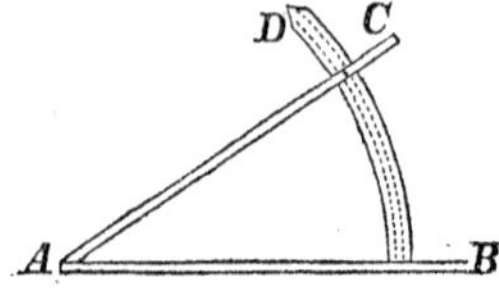


5. Wishing to know the perpendicular hight of the hill, I constructed an instrument consisting of two legs, which revolved on a point, as shown in the accompanying figure. The instrument could be set at any desirable angle, by pinning the leg *AC* to the arc *D*. The legs were each 10 feet long. On placing the leg *AB* in a horizontal position, and elevating the end *C* of the leg *AC*, till, with my eye at *A*, the upper edge of the leg was in range with the foot of the tree, I found the perpendicular distance from *C* to the leg *AB* to be just 2 feet; allowing the leg *AB* to be 5 feet above the ground, what was the perpendicular hight of the hill? *Ans.* 389 feet.

6. On the same hill, at the same distance from me, was a rock, and I wished to ascertain its distance from the tree. To do this, I used the same instrument, with both its legs lying in the same plane. I placed the point *A* over the first point of observation, the leg *AB* in range with the foot of the tree, and the leg *AC* in range with the base of the rock. Then the distance from *B* to *C* was 4 inches; what was the distance from the tree to the rock? *Ans.* 64 feet.

## ¶ 45. *To find the area of a right-angled triangle.*

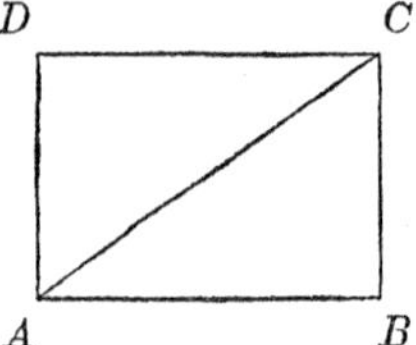


Analysis. By an inspection of the accompanying diagram, it will readily be seen that the rectangle *ABCD* contains two right-angled triangles, *ABC* and *ADC*; consequently the area of any right-angled triangle is equal to one half the area of a rectangle having the same base and perpendicular. Hence,

### *To find the area of a right-angled triangle.*

**RULE.**

Multiply one half the base by the perpendicular, or one half the perpendicular by the base. *Or,*

Multiply the base by the perpendicular, and take one half of the product.*

Note 1. 625 sq. L., or 272·25 sq. ft., = 1 P., and 16 P. = 1 A. (See ¶¶ 15 and 16.) When the area of a triangle is in sq. L., it will be reduced to P. by

¶ **45.** Topic. Analysis. Rule. Note 1. Reference to second analysis. Ex. 4, note 3.

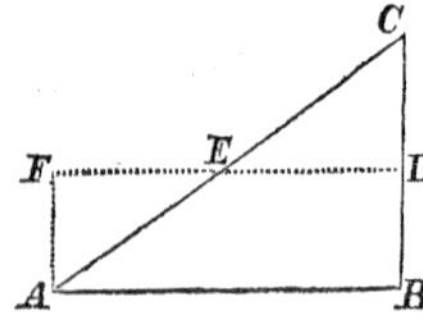


* The principles of this rule may also be analyzed as follows:—

If from the triangle *ABC* a portion be cut off parallel to the base, as from *D* to *E*, bisecting the perpendicular and hypotenuse, the portion *CDE* will be equal to the triangle *AFE*, and, if placed there, will complete the rectangle *ABDF*, which is equal to the triangle *ABC*.

dividing by 625; when in sq. ft., by dividing by 272'25; and when in P., it will be reduced to A. by dividing by 16. But the base multiplied by the perpendicular gives twice the area of the triangle; and this area, divided by twice the number that it takes of that denomination in which the area is expressed to make one of the next higher denomination, will reduce the area to the next higher denomination. 625 sq. L. $\times$ 2 = 1250 sq. L.; 272'25 sq. ft. $\times$ 2 = 544'5 sq. ft.; and 16 P. $\times$ 2 = 32 P. Hence, *To find the number of acres in a triangle, multiply the base by the perpendicular. If the dimensions are given in links, divide the product by 1250, and if in feet, by 544'5, and the quotient will be poles, which may be reduced to acres. If the dimensions are given in poles, divide the product by 32, and the quotient will be acres.* 1250 is $\frac{1}{8}$ of 10000, $\frac{1250}{10000} = \frac{1}{8}$. Multiplying by 8 and dividing the product by 10000, is the same in effect as dividing by 1250. Therefore, instead of dividing by 1250, we may multiply by 8, and cut off 4 figures from the right hand of the product.

## EXAMPLES FOR PRACTICE.

1. What are the contents of a right-angled triangle whose base is 12 inches, and perpendicular 8 inches?

2. The legs of a right-angled triangle are 70 rods, and 40 rods in length; what are the contents of the triangle in acres? *Ans.* 8¾ acres.

3. The gable ends of a barn are each 28 feet wide, and the perpendicular hight of the ridge of the roof above the eaves is 7 feet; how many feet of boards will be required to board up the gables? *Ans.* 196 feet.

NOTE 2. By reference to ¶ 39, ex. 2, it will be seen that the gable may readily be resolved into two right-angled triangles.

4. The area of a right-angled triangle is 48 feet, and the base is 12 feet; what is the perpendicular?

NOTE 3. This example is the reverse of the preceding ones in this ¶, and may be performed by reversing the rule.

5. The area of a right-angled triangle is 72 yards, and the perpendicular is 9 yards; what is the base? *Ans.* 16 yards.

6. The area of the gable of a certain building is 108 feet, and the perpendicular hight of the ridge of the roof above the eaves is 9 feet; what is the width of the building? *Ans.* 24 feet.

### ¶ 46. *To find the area of an equilateral and of an isosceles triangle.*

ANALYSIS. Any equilateral triangle, or any isosceles triangle, may be resolved into two right-angled triangles, each having for its base one half the base of the given triangle, and both having one common perpendicular. Thus, the equilateral triangle $ABC$ consists of the two right-angled triangles $ADC$ and $BDC$. Hence,

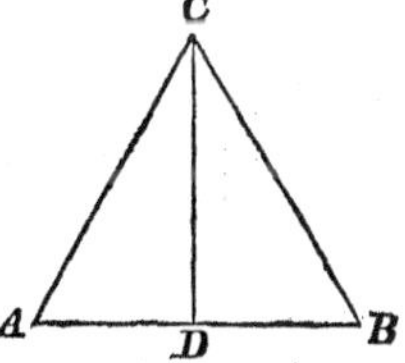


¶ 46. Topic. Analysis. Conclusion. Note.

*The area of an equilateral, or of an isosceles triangle may be found by the principles of the rule* ¶ 45.

**EXAMPLES FOR PRACTICE.**

1. What are the contents of a triangle whose sides measure 18 inches each? *Ans.* 280'44+ sq. in.

2. What is the area of an equilateral triangle whose side measures 20 yards? *Ans.* 346'4+ sq. yds.

3. What is the area of an isosceles triangle, the base or longest side of which is 16 feet, and the other sides are each 13 feet 4 inches? *Ans.* 85 sq. ft. 4 sq. in.

4. Two sides of a field, in the form of an isosceles triangle, are each 60 rods long, and the other side is 96 rods long; how many acres does the field contain? See ¶ 45, Note 1. *Ans.* 10 A. 3 R. 8 P.

NOTE. When the length of one side of any triangle, and the perpendicular distance between this side and the opposite angle, are given, the area may be found by an application of the same principles.

5. The base of a triangle is 18 inches, and the perpendicular 13 inches; what is the area?

6. One side of a field, in a triangular form, is 18 chains in length, and the perpendicular distance between this side and the opposite angle is 15 chains; what is the area of the field? *Ans.* 3 R. 15 P.

7. One side of the roof of a building is 16 feet wide, the other side 18, and the perpendicular hight of the ridge above the eaves is 9 feet; how many clapboards, each covering 4 inches wide by 13 feet long, will be required to cover both gables? *Ans.* Nearly 60.

**¶ 47.** *To find the area of any triangle, the length of the sides being given*

FIRST METHOD.

ANALYSIS. We may construct any number of triangles, having the same base and a common altitude, as *ABC*, *ABD*, *ABE*, &c., and divide them into right-angled triangles, by ¶ 35, Prob. VII., and their areas will all be the same. Consequently, all triangles constructed on the same base, and having the same altitude, are equal. Hence,

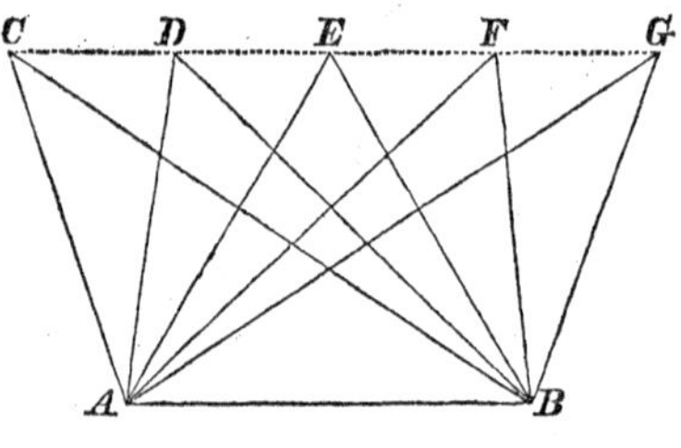


**RULE.**

I. Construct the given triangle, as taught in ¶ 35, Prob. XIV.

---

**¶ 47.** Topic. Analysis of first method. Rule. Second rule. Note.

II. On one side of the triangle erect a perpendicular equal to the altitude of the triangle.

III. Multiply the base by the perpendicular, and take one half of the product for the area. See ¶ 45.

### SECOND METHOD.

**RULE.**

I. Add the lengths of the three sides together, and from half their sum subtract the length of each side separately.

II. Then multiply the half length of the three sides and the three remainders together, and extract the square root of their product.

NOTE. The principles of this rule do not admit of an arithmetical analysis.

**EXAMPLES FOR PRACTICE.**

1. How many poles in a triangular field, the sides of which measure respectively 13, 14, and 15 rods? *Ans.* 84.

2. How many square inches in a triangular board whose sides measure respectively 22, 26, and 30 inches? *Ans.* 278'51+.

3. A triangular field. whose sides measure 386, 420, and 765 yards leases annually for $1'87½ per acre; how much is the annual rent? *Ans.* $17'316+.

## ¶ 48. *To find the area of a rhombus and of a rhomboid.*

ANALYSIS. If the right-angled triangle *AED* be placed on the opposite side of the rhombus *ABCD*, it will fill the space *BFC*, and the rhombus will then be reduced to a square. By the same process the rhomboid will be reduced to a rectangle. Hence,

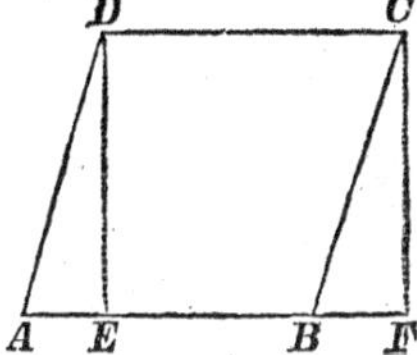


*To find the area of a rhombus or of a rhomboid.*

**RULE.**

Multiply the length by the shortest or perpendicular distance between two opposite sides.

**EXAMPLES FOR PRACTICE.**

1. The side of a rhombus is 18 inches long, and the shortest distance between its opposite sides is 14 inches; what is its area?

2. A meadow, in the form of a rhomboid, is 20 chains long, and the shortest distance between its opposite sides is 12 chains: how many hours will it take a man to mow the grass on this meadow, if he mow 1 square rod in 3 minutes? How many days, if he work 10 hours each day? *Ans.* 19 da. 2 h.

---

¶ 48. Topic. Analysis. Rule.

3. The side of a board, in the form of a rhombus, is 15 inches long, and a perpendicular, running from one obtuse angle, will meet the opposite side 9 inches from the acute angle; what is the length of the perpendicular, and what is the area of the board? *Ans.* to last. 180 sq. in.

## ¶ 49. *To find the area of a trapezoid.*

The side $AB$ is 24 inches, the side $CD$ 16 inches, and the altitude or distance $ad$ 7 inches; what is the area of the trapezoid?

ANALYSIS. If the triangle $Aae$ be applied to the space $Dde$, and the triangle $Bbe$ to the space $Cce$, the trapezoid will be reduced to a rectangle, the side $ab$ being equal to the side $cd$. The side $CD$ will be increased just as much as the side $AB$ is diminished, and the sides $ab$ and $cd$ will each be equal to one half the sum of the sides $AB$, $CD$. $24 + 16 = 40$, $40 \div 2 = 20$, and $20 \times 7 = 140$, the number of square inches in the trapezoid. Hence,

### *To find the area of a trapezoid.*

**RULE.**

Multiply one half the sum of the parallel sides by the perpendicular distance between them.

**EXAMPLES FOR PRACTICE.**

1. What are the square contents of a board 12 feet long, 16 inches wide at one end, and 9 at the other? *Ans.* 12½ sq. ft.
2. What are the contents of a stock of 12 boards, 14 feet long, 10 inches wide at one end, and 2½ at the other? *Ans.* 87½ sq. ft.
3. What is the area of a board 12 feet long, 16 inches wide at each end, and 8 in the middle? *Ans.* 12 sq. ft.
4. One side of a field is 40 chains long, the side parallel to it is 22 chains long, and the perpendicular distance between these two sides is 25 chains; how many acres in the field? *Ans.* 77 A. 5 sq. C.

## ¶ 50. *To find the area of a trapezium.*

ANALYSIS. If from any angle a diagonal be drawn to the opposite angle, as $AC$, the figure will be divided into two triangles, $ABC$ and $ACD$. Consequently, if the sides and a diagonal of a trapezium be given, the sides of the triangles which compose it are also given. Hence,

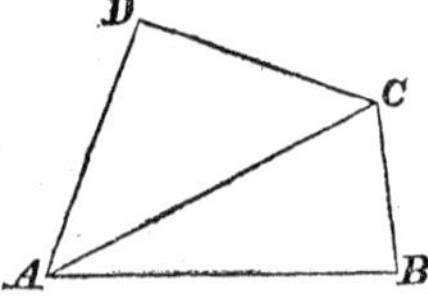


¶ 49. Topic. Analysis. Rule.
¶ 50. Topic. Analysis. Rule.

*To find the area of a trapezium.*

**RULE.**

I. Divide the trapezium into two triangles, by drawing a diagonal to two opposite angles, and from each of the other two angles let fall a perpendicular to the diagonal.

II. Multiply the sum of the perpendiculars by one half the diagonal, or the diagonal by one half the sum of the perpendiculars. *Or*,

Find the area of each triangle separately, ¶ 47, and add them together.

**EXAMPLES FOR PRACTICE.**

1. The diagonal of a trapezium is 25 feet, and the perpendiculars are 9 and 13 feet; what is the area? *Ans.* 275 sq. ft.

2. The diagonal of a trapezium is 34 feet 9 inches, and the sum of the perpendiculars is 28 ft. 6 inches; what is the area? *Ans.* 495 ft. 2′ 3″.

3. The diagonal of a trapezium is 32 yards, the two remaining sides of one triangle are 20 and 26 yards, and the two remaining sides of the other are 29 and 9 yards; what is the area of the trapezium?
*Ans.* 387‘65+ sq. yds.

4. The diagonal of a field, in the form of a trapezium, is 538 yards, the two remaining sides of one triangle are 283 and 471 yards, and of the other 432 and 216 yards; how many acres in the field?
*Ans.* 22 A. 3 R. 31 P. 18‘66 sq. yds.

## ¶ 51. *Similar rectilinear figures.*

1. What ratio does a figure 4 inches square bear to one 2 inches square? ANALYSIS. $4^2=16$, $2^2=4$, and $16\div4=4$.
*Ans.* The ratio of 4 to 1.

2. How does the area of a square foot compare with that of a square yard? *Ans.* It is $\frac{1}{9}$ as large.

3. I have 4 small fields; the largest is 40 rods square, and the other 3 are each 20 rods square; how does the largest field compare in size with one of the others? How with all of them? *Ans.* to last. As 4 to 3.

4. How many boards, 3 feet long and 2 feet wide, will be required to cover a space 9 feet long and 6 feet wide?

5. Two rectangles measure as follows: the smaller 5 by 12 inches, and the larger 20 by 48 inches; how many of the smaller ones will be equal to the larger one? *Ans.* 16.

6. The two shorter sides of one right-angled triangle measure 8 and 15 feet, and the two shorter sides of another 16 and 30 feet; how do the areas compare with each other? *Ans.* As 1 to 4.

7. The 3 sides of one triangle measure respectively 18, 14, and 10 inches, and the 3 sides of another 9, 7, and 5 inches; what ratio does the area of the first bear to that of the second? *Ans.* The ratio of 4 to 1.

8. The side of a regular octagon measures 9 inches, and the side of a

---

**¶ 51.** Topic. Object of Ex. 1. — of Ex. 2. — Ex. 4. — Ex. 6. — Ex. 8.

second similar figure measures 45 inches; the second is how many times as large as the first? How many times larger?

*Ans.* to last. 24 times larger.

Ex. 1 shows that a square whose sides are double the length of another square, contains 4 times the area.

Ex. 2 shows that a square whose side is 3 times as long as the side of another square, contains 9 times the area.

Ex. 4 shows that a rectangle twice as long and twice as wide as another rectangle, contains 4 times the area.

Ex. 6 shows that any triangle whose sides measure twice as much as the sides of another similar triangle, contains 4 times the area.

Ex. 8 shows that any regular polygon whose side is 5 times as long as the side of another similar polygon, contains 25 times the area.

In each of the preceding examples in this ¶, the area of the larger figure can be obtained, by multiplying the area of the smaller figure by the square of the number of times the side of the smaller figure is contained in the side of the larger, and *vice versa.* Hence,

**¶ 52.** *The areas of similar rectilinear figures are to each other as the squares of their similar sides;* and

*The similar sides are to each other as the square root of the quotient of the area of the greater figure divided by the area of the less.*

NOTE. The pupil will find by trial that the second of the above principles is the reverse of the first; and that, having the areas of two similar figures, and any side of one of them given, the similar side of the other, and the ratio between all the similar sides of the two, may be found, by an application of these principles.

**¶ 53.** *To find the area of any regular polygon.*

ANALYSIS. Any regular polygon may be resolved into as many equal triangles as the polygon contains sides. The base of each triangle will be the length of one side of the polygon, and the altitude of each will be the perpendicular distance from the center of the polygon to the middle of one side. Hence,

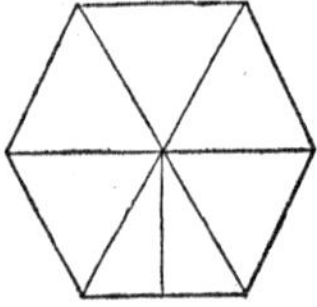

---

**¶ 52.** First principle. Second. Note.
**¶ 53.** Topic. Analysis. Rule. Note.

*To find the area of any regular polygon.*

RULE.

I. Bisect any two contiguous sides of the polygon, the point where the bisecting lines intersect each other will be the center of the polygon, or the apex of the triangle of which the polygon is composed; and the perpendicular distance from the center to one side will be the altitude of each triangle.

II. Multiply the length of the perpendicular by one half the perimeter.

NOTE. We may first find the area of one triangle, ¶ 47, and then multiply this area by the number of triangles in the given polygon.

EXAMPLES FOR PRACTICE.

1. What is the area of a regular pentagon, the side of which measures 25 inches, and the perpendicular distance from the center to the middle of one side is 17‘2 inches? *Ans.* 7 sq. ft. 67 sq. in.

2. What is the area of a regular hexagon, the side of which is 10 feet, and the altitude of one of its equal triangles is 8‘660254 feet? *Ans.* 259‘80762 sq. ft.

**¶ 54.** The areas of similar polygons are to each other as the squares of one of their sides. ¶ 52. Hence, the areas of regular polygons may be more readily found by the help of a table prepared in the following manner.

Consider the sides of each of the regular polygons to be 1; then find the perpendicular and area.

| Number of sides. | Names of figures. | Areas and Multipliers. |
|---|---|---|
| 3 | Triangle, | ‘433013 |
| 4 | Square, | 1‘000000 |
| 5 | Pentagon, | 1‘720477 |
| 6 | Hexagon, | 2‘598076 |
| 7 | Heptagon, | 3‘633912 |
| 8 | Octagon, | 4‘828427 |
| 9 | Nonagon, | 6‘181824 |
| 10 | Decagon, | 7‘694209 |
| 11 | Undecagon, | 9‘365640 |
| 12 | Dodecagon, | 11‘196152 |

The above table shows the area of polygons of any number of sides from 3 to 12, each side being unity or 1. Hence,

*The length of one side, and the number of sides of any regular polygon being given, to find the area, by the above table.*

---

¶ 54. Principle. Reference. Formation of table. Its use. Rule. Note.

**RULE.**

Multiply the square of the given side by the tabular number of a similar polygon.

NOTE. If the area, the number of sides, and the length of one side be given, the perpendicular may be found by reversing this rule. Also, if the area, the number of sides, and the perpendicular be given, the length of one side may be found by reversing the rule. Also, the area, the perpendicular, and the length of one side being given, the number of sides may be found by a reverse process.

**EXAMPLES FOR PRACTICE.**

1. What is the area of a triangle whose side is 15 inches?
*Ans.* 97'4+ sq. in.
2. What is the area of a square whose side is 5 miles?
3. What is the area of a pentagon whose side is 8 feet?
*Ans.* 110 sq. ft. 15'9+ sq. in.
4. What is the area of a hexagon whose side is $2\frac{1}{2}$ yards?
*Ans.* 16 sq. yds. 2 sq. ft. 20'4+ sq. in.
5. What is the area of a heptagon whose side is $\frac{1}{4}$ of a rod?
*Ans.* 6 sq. yds. 7 sq. ft. 120 sq. in.
6. The side of an octagon is $4\frac{1}{3}$ feet; what is the area?
*Ans.* 90'667+ sq. ft.
7. The side of a nonagon is $17\frac{1}{3}$ inches; what is the area?
*Ans.* 12'897+ sq. ft.
8. The side of decagon is $3\frac{1}{5}$ yards; what is the area?
*Ans.* 78 sq. yds. 7 sq. ft. 14+ sq. in.
9. The side of an undecagon is 46 chains and 15 links; what is the area?
*Ans.* 1994 A. 7'15+ sq. C.
10. The side of a dodecagon is $\frac{3}{8}$ of an inch; what is the area?
*Ans.* 1'57+ sq. in.

**¶ 55.** *To find the area of any irregular rectilinear figure or polygon.*

**RULE.**

I. Divide the figure into as many triangles as may be, by drawing diagonals from any one angle to all the others.

II. Find the areas of the several triangles, and add them together.

NOTE 1. Any rectilinear figure may be divided into as many triangles, less two, without any of the dividing lines crossing each other, as the figure has sides.

NOTE 2. The area may frequently be more easily found, by dividing the polygon into as many trapezia as may be, thus diminishing the number of triangles; for the area of a trapezium is more readily obtained than the areas of the two triangles which compose it.

---

¶ **55.** Topic. Rule. Note 1. Note 2.

**EXAMPLES FOR PRACTICE.**

1. A piece of land of an irregular figure, bounded by right lines, is divided into 3 trapezia: in the first the diagonal is 4 C. 24 l., and the sum of the perpendiculars 3 C. 67 l.; in the second the diagonal is 7 C. 43 l., and the sum of the perpendiculars 5 C. 38 l.; and in the third the diagonal is 6 C. 78 l., and the sum of the perpendiculars 4 C. 84 l.; how many acres does the field contain? *Ans.* 4'41747 A.

---

## Board or Lumber Measure.

**¶ 56.** The standard of thickness for boards is 1 inch. All lumber not exceeding 1 inch in thickness is bought and sold by the superficial measure of 144 square inches to the foot. All lumber exceeding 1 inch in thickness is first reduced to the standard thickness, and then estimated by superficial measure, as before.

**¶ 57.** *To find the number of feet in a straight-edged board, of uniform width.*

For rule and principles consult ¶ 37.

*To find the number of feet in a board that tapers.*

For rule and principles consult ¶ 49.

Note 1. The mean width of a tapering board is usually taken, by measuring the width of the board at an equal distance from each end.

**EXAMPLES FOR PRACTICE.**

1. How many feet in a board 58 inches long and 9 inches wide?

Note 2. When the length and width are given in inches, divide the product by 144 to reduce to square feet. Why?

2. How many feet in a board 18 feet long and 15 inches wide?

Note 3. When the length is given in feet, and the width in inches, divide the product by 12 to reduce to square feet. Why? *Ans.* 22½.

3. How many feet in 4 boards, each 12 ft. 7 in. long, and 1 ft. 4 in. wide?

Note. 4. The inches may be reduced to fractions of a foot, and the whole to improper fractions, and multiplied; the feet may be reduced to inches, and then multiplied; or, the multiplication may be performed by duodecimals. See Revised Arithmetic, ¶ 204.

---

**¶ 56.** Standard thickness of boards. Lumber less than 1 inch thick. Lumber more than 1 inch thick.

**¶ 57.** Number of feet in a straight-edged board of uniform width; — in a straight-edged board that tapers. Mean width of a tapering board. Both dimensions in feet; — in inches. One dimension in feet and the other in inches. Dimensions in feet and inches.

4. How many feet in a stock of 9 boards, each 13 feet long, 9 inches wide, and $1\frac{1}{2}$ inches thick? *Ans.* $131\frac{5}{8}$.

5. How many feet in a stock of 5 boards, each 11 feet long, 15 inches wide at one end, and 12 inches at the other?

6. How many feet in a stock of 13 boards, each 14 feet long, 16 inches wide at one end, and running to a point at the other? *Ans.* $121\frac{1}{3}$.

7. What length of board 9 inches wide will make a square foot?

8. What length of board 15 inches wide will be required to make a fireboard 3 feet square?

9. How many feet, board measure, in a stock of 7 planks, each 12 feet long, 22 inches wide, and $2\frac{1}{4}$ inches thick? *Ans.* $346\frac{1}{2}$.

10. What are the contents of a stock of 9 boards, 13 feet long, 18 inches wide at each end, and 14 in the middle?

11. What are the contents of a stock of 7 boards, 13 feet long, 10 inches wide at one end, 12 at the other, and 8 at the distance of 7 feet from the wider end? *Ans.* 72 sq. ft. 48 sq. in.

**¶ 58.** *To find the number of feet of straight-edged boards in a stock of wane-edged boards, sawn from a round log, without slabbing.*

In sawing a round log into wane-edged boards, a slab of 1 inch is taken off from two opposite sides, and the remainder of the log is sawn into boards, which of course are of different widths. From an actual measurement of stocks of boards, sawn from logs in the manner above described, the following facts have been deduced.

1. *In stocks of boards sawn from logs from* 7 *to* 12 *inches in diameter, the width of the second board is the average width of the whole stock.*

2. *In stocks of boards sawn from logs from* 12 *to* 24 *inches in diameter, the width of the third board is the average width of the whole stock.*

3. *In stocks of boards sawn from logs from* 24 *to* 36 *inches in diameter, the width of the fourth board is the average width of the whole stock.*

In all cases a slab of 1 inch in thickness is to be taken off, before the first board is sawn; and the board is to be measured on the narrower side; *i. e.*, the side towards the slab. The above averages will vary slightly, but will not make a difference of more than $2\frac{1}{2}$ feet in any log from 7 to 36 inches; the difference being sometimes in favor of the buyer, and at others in favor of the seller.

---

**¶ 59.** *The diameter of a circle being given, to find the circumference.*

---

¶ **58.** Sawing wane-edged boards. First fact. Second. Third. Slabbing. Measuring the average board. Variation of these averages.

It has been a problem among the ablest mathematicians, for ages, to find the exact ratio between the diameter and circumference of a circle. Although this ratio has never yet been definitely ascertained, yet results have been arrived at, which are sufficiently exact for all practical purposes.

It has been found that the diameter of a circle is to the circumference *nearly* as 1 to $3\frac{1}{7}$, or 7 to 22. A nearer approximation is as 113 to 355, or 1 to 3·14159. The former ratio is sufficiently exact for ordinary mechanical purposes, but in estimating machinery, and other calculations where greater accuracy is required, the latter ratio should be used. Hence,

*When the diameter of a circle is given, to find the circumference.*

**RULE.**

Multiply the diameter by $3\frac{1}{7}$; *or*, where greater accuracy is required, by 3·14159. *Or*, for ordinary purposes, say,

7 : 22 : : the given diameter : the required circumference. And, where greater accuracy is required, 113 : 355 : : the diameter : the circumference.

NOTE 1. The latter ratio has the advantage of being easily remembered, the numbers being formed of the first three odd numbers, each repeated.

NOTE 2. Most of the operations under this head, in this work, are performed by the proportion 113 : 355 : : the given diameter : the circumference.

**EXAMPLES FOR PRACTICE.**

1. What is the circumference of a circle 8 feet in diameter? *Ans.* 25 ft. 1·58+ in.

2. What is the circumference of a circle 7 inches in diameter, by the first proportion? — by the second? Which gives the greater result, and how much?
*Ans.* The first gives ·00885 of an inch greater than the second.

3. The diameter of a certain wheel is 10·5 feet; what is its circumference? *Ans.* 32 ft. 11·83+ in.

4. What length of tire will it take to band a carriage-wheel 5 feet in diameter?

5. What is the circumference of a circle 113 rods in diameter, by the first proportion? — by the second? Which gives the greater result, and how much?

6. What is the circumference of a circular lake 721 rods in diameter? *Ans.* 7 mi. 25 rds. 1·45+ ft.

7. A horse is made fast to a stake by a line, one end of which is fastened to his nose, and the other to the stake; allowing this line to be 25 feet long, what is the circumference of the circle upon which he may feed? *Ans.* 157·07+ ft.

---

¶ 59. Topic. Analysis. Rule. Note 1. Note 2.

**¶ 60.** *The circumference of a circle being given, to find the diameter.*

**RULE.**

Divide the circumference by $3\frac{1}{7}$; *or*, where greater accuracy is required, by 3'14159.

*Or*, for ordinary purposes, say 22 : 7 : : circumference : diameter. And where greater accuracy is required, 355 : 113 : : circumference : diameter.

NOTE. Since this rule is the reverse of the rule ¶ 59, no analysis is deemed necessary.

**EXAMPLES FOR PRACTICE.**

1 What is the diameter of a circle 33 yards in circumference?

2. If the circumference be 49'52 rods, what is the diameter? *Ans.* 15'762+ rds.

3. The circumference of a cart-wheel is 16 feet 6 inches; what is the diameter?

4. A circular park is 320 links in circumference; what is its diameter?

5. If the extreme end of the minute-hand of a clock move forward 19 inches in 12 minutes, what is the circumference of the dial-plate? What is the length of the minute-hand? *Ans.* to last. $15\frac{17}{142}$ inches.

6. Within a circular garden 66 chains in circumference, is a circular pond 66 rods in circumference; what is the diameter of the garden?— of the pond? *Ans.* { Diameter of garden, 21 C. 6'69 in. " " pond, 21 rds. 1'67 in.

**¶ 61.** *The number of degrees in a circular arc, and the radius of the circle being given, to find the length of the arc.*

**RULE.**

I. Find the circumference of the circle.

II. Then say, 360° : the number of ° in the arc : : the circumference of the circle : the length of the arc.

NOTE. The reasons for the rule are obvious.

**EXAMPLES FOR PRACTICE.**

1. What is the length of an arc of 18°, in a circle whose radius is 4 ft. 8'5 in.? *Ans.* 1 ft. 5'75 in.

2. What is the length of an arc of 36° 15', in a circle whose radius is 15 feet?

3. What is the length of an arc of 225° 7' 30", in a circle 12 rods in diameter? *Ans.* 7 rds. 8 ft. 3'72+ in.

4. The length of an arc is $17\frac{3}{4}$ inches, and the remaining part of the circle contains 342°; what is the diameter of the circle? *Ans.* 29 ft. 7 in.

---

**¶ 60.** Topic. Analysis. Rule.
**¶ 61.** Topic. Analysis. Rule.

## ¶ 62. *To find the area of a circle.*

Analysis. Any circle may be supposed to be divided into an infinite number of equal isosceles triangles, whose apexes all meet in the center of the circle, and whose bases all lie in the circumference. In other words, the circle may be considered as a regular polygon of an infinite number of sides, the perimeter of the polygon being the circumference of the circle, and the altitude of one of the equal triangles of which it is composed being the radius. And, since the area of a regular polygon is found by multiplying the perpendicular by one half the perimeter, (¶ 53, rule,) the area of a circle may be found by applying the same principles. Hence,

### *To find the area of a circle.*

**RULE.**

Multiply one half the circumference by one half the diameter.

Note 1. If the whole circumference be multiplied by the whole diameter, and the product divided by 4, the quotient will be the area. This operation will in many cases obviate the use of fractions in multiplying and dividing.

**EXAMPLES FOR PRACTICE.**

1. The diameter of a circle is 7, and the circumference 22; what is the area?
2. The diameter of a circle is 113, and the circumference 355; what is the area?
3. What is the area of a barrel head 16 inches in diameter?

Note 2. First find the circumference, by ¶ 59, rule.

*Ans.* 201'04 sq. in.

4. What is the area of a circle described with a radius 2 chains in length?
5. The circumference of the end of a log is 82 inches; what is the area? *Ans.* 3 sq. ft. 103'05+ sq. in.

## ¶ 63. *To find the area of a circle, when the diameter only is given.*

Ex. What is the area of a circle inscribed within a figure 1 foot square?

Analysis. The diameter of the circle is equal to one side of the square, or 1 foot. 1 ft. (diameter) ÷ 2 = '5 foot; 3'14159 ft. (circumference) ÷ 2 = 1'570795 ft. Then, '5 × 1'570795 = '7853975, the area of any circle whose diameter is 1. This area wants but '0000025 = $\frac{1}{400000}$ of being '7854, which is so small a fraction, that in business calculations it is disregarded, and the area of a circle is estimated to be '7854 of the area of its superscribing square. Hence,

---

¶ 62. Topic. Analysis. Rule. Note 1.
¶ 63. Topic. Solution of Ex. Conclusion deduced from solution. Rule.

*To find the area of a circle, when the diameter only is given.*

**RULE.**

Multiply the square of the diameter by '7854.

**EXAMPLES FOR PRACTICE.**

1. What is the area of a circle 7 inches in diameter?
2. What is the area of one surface of a circular saw 25 inches in diameter? *Ans.* 3 sq. ft. 58$\frac{7}{8}$ sq. in.
3. What is the area of a circle $\frac{1}{4}$ of an inch in diameter?
4. How many circles 1 inch in diameter, are equal to a circle 4 inches in diameter? *Ans.* 16.

## ¶ 64. *The area of a circle being given, to find the diameter.*

**RULE.**

Divide the area by '7854, and extract the square root of the quotient.

NOTE 1. This rule being the reverse of the rule ¶63, an analysis is deemed unnecessary.

**EXAMPLES FOR PRACTICE.**

1. What is the diameter of a circle whose area is 38$\frac{1}{2}$ square feet? *Ans.* 7 ft. '01+ in.
2. What is the diameter of a circular park which contains 2464 square yards?
3. The area of a circle is 78'54 chains; what is the circumference?

NOTE 2. First find the diameter.

*Ans.* 31 C. 41'59+ l.

4. What is the diameter of a circular island containing 1 square mile of land? *Ans.* 1 mi. 41 rds. 5'83+ ft.
5. What is the circumference of a circular pond which covers 7'0686 square chains?

## ¶ 65. *To find the area of a semicircle, a quadrant, and a sextant.*

ANALYSIS. The area of a semicircle is equal to $\frac{1}{2}$, the area of a quadrant to $\frac{1}{4}$, and the area of a sextant to $\frac{1}{6}$ the area of a circle having the same radius. Hence,

*To find the area of a semicircle, a quadrant and a sextant.*

---

¶ **64.** Topic. Analysis. Rule.

¶ **65.** Topic. Analysis. Rule for the semicircle; — for the quadrant; — for the sextant. Note 1. Reasons for these operations.

### RULE.

I. *For the semicircle;* — Take $\frac{1}{2}$ of the area of a circle having the same radius.

II. *For the quadrant;* — Take $\frac{1}{4}$ the area of a circle having the same radius.

III. *For the sextant;* — Take $\frac{1}{6}$ the area of a circle having the same radius.

NOTE 1. The area of the semicircle may be found, by multiplying $\frac{1}{2}$ the square of twice the radius by '7854, or twice the square of the radius by '7854. *Also*, the area of the quadrant may be found, by multiplying $\frac{1}{4}$ the square of twice the radius by '7854, or the square of the radius by '7854. The intelligent pupil will readily deduce these principles from ¶ 63, rule, and from the principles of simple multiplication.

### EXAMPLES FOR PRACTICE.

1. What is the area of $\frac{1}{2}$ of a circle 16 inches in diameter?
2. What is the area of a semicircle whose radius is 18 feet?
3. What is the area of a quadrant whose radius is 4 yards?
4. What is the area of a sextant whose radius is 11 inches?
5. What is the radius of a quadrant whose area is 12'5664 square yards? *Ans.* 12 ft.

NOTE 2. This example is the reverse of Ex. 3. The pupil will deduce his principles for operation from ¶ 45, Note 3.

6. The area of a semicircle is 3'5343 square chains; what is the radius? *Ans.* 18 rds.
7. The area of a sextant is 63'3556 square inches; what is the radius?

## ¶ 66. *To find the area of a sector, the radius and arc being given.*

ANALYSIS. Since the whole circumference of a circle contains the whole area, any arc and its radii will contain such a part of the whole area, as the arc is part of the whole circumference. *Or,*

A sector may be supposed to consist of an infinite number of triangles, the same as the circle in ¶ 62. Hence,

*To find the area of a sector, the radius and arc being given.*

### RULE.

First, find the circumference of the original circle, and also its area. Then say,

The whole circumference : the given arc : : the whole area the area of the sector. *Or,*

Multiply the length of the arc by $\frac{1}{2}$ the radius.

### EXAMPLES FOR PRACTICE.

1. The radius of a sector is 15 inches, and the arc $4\frac{1}{2}$ inches; what is the area?

---

¶ 66. Topic. Analysis. Rule.

2. The radius is 48 yards, and the arc 50 yards; what is the area of the sector?

3. What is the area of a sector whose radius is 200 rods, and arc 12½ feet?

4. The area of a sector is 33¾ inches, and the radius 15 inches; what is the arc?

NOTE. This example is the reverse of Ex. 1.

**¶ 67.** *To find the area of a sector, the radius and the angle at the center being given.*

ANALYSIS. Since the whole circumference of a circle contains 360°, the sum of all the angles that can be made, by radii drawn from the circumference to the center of any circle, must be 360°; and any sector must contain such a part of the area of the whole circle, as the number of degrees contained in its angle at the center is part of 360°. Hence,

*To find the area of a sector, the radius and angle at the center being given.*

### RULE.

First find the area of the original circle. Then say,

360° : the given angle :: the area of the original circle : the area of the sector.

### EXAMPLES FOR PRACTICE.

1. The radius of a sector is 7 feet, and the angle at the center 45°; what is the area of the sector? *Ans.* 19'2423 sq. ft.

2. The radius of a sector is 24 ft. 6 in., and the arc contains 137° 30′; what is the area of the sector? *Ans.* 196 sq. ft. 62'181 sq. in.

3. The radius of a sector is 113 inches, and the area 17 sq. ft. 59'19315 sq. in.; what is the angle at the center? *Ans.* 22° 30′.

NOTE. This example is the reverse of the two preceding.

**¶ 68.** *To find the side of a square which shall contain an area equal to a given circle.*

### RULE.

Find the area of the given circle, and extract its square root.

NOTE. The pupil will readily analyze this rule.

### EXAMPLES FOR PRACTICE.

1. What is the side of a square equal in area to a circle 14 feet in diameter? *Ans.* 12'407+ ft.

---

**¶ 67.** Topic. Analysis. Rule.
**¶ 68.** Topic. Analysis. Rule.

2. I have a circular garden 42 rods in circumference, and I wish to lay out a square park of the same area as the garden; what will be the length of one side of the park?

3. One monument is built upon a circular base, and another, which stands near it, upon a square base, the area of the base of the latter being equal to that of the former. The side of the base of the square monument is 5 feet; what is the diameter of the base of the circular monument? *Ans.* 5'641+ ft

## ¶ 69. *The diameter of a circle being given, to find the side of its inscribed equilateral triangle.*

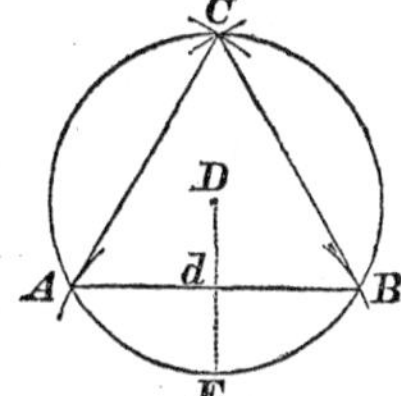


Analysis. The side *AB*, of the equilateral triangle *ABC*, intersects the radius *DE* of the superscribing circle, at the point *d*, equi-distant from *D* and *E*. Consequently, the side *AB* is the base of two triangles, *AdE* and *BdE*, the hypotenuse of each being the radius of the circle, or the side of the inscribed hexagon, and the perpendicular of each ½ the radius *DE*. We now have the hypotenuse *AE* or *BE*, (equal to the radius of the given circle,) and the perpendicular *Ed*, (equal to ½ the radius,) to find the base *Ad* or *Bd*, which, being multiplied by 2, will be the side of the inscribed equilateral triangle. Hence,

*To find the side of an equilateral triangle inscribed in a given circle.*

### RULE.

I. From the square of the radius subtract the square of half the radius.

II. Extract the square root of the remainder, and multiply the result by 2.

Note 1. Instead of the second operation directed in the rule, the same result will be obtained by multiplying the remainder by 4, and extracting the square root of this product.

### EXAMPLES FOR PRACTICE.

1. The radius of a circle is 8 inches; what is the side of an inscribed equilateral triangle? *Ans.* 13'85+ in.

2. What is the side of an equilateral triangle inscribed in a circle 50 yards in diameter? *Ans.* 86 yds. 1 ft. 9'69+ in.

3. What is the side of the greatest equilateral triangle that can be cut from a circular plate of copper 3 inches in diameter? *Ans.* 2'598+ in.

4. The side of an equilateral triangle is 14 inches; what is the diameter of the superscribing circle?

Note 2. This example is the reverse of the three preceding ones. *Ans.* 16'16+ in.

---

¶ 69. Topic. Analysis. Rule. Note 1. Method of performing Ex. 4.

**¶ 70.** *The diameter of a circle being given, to find the side of its inscribed square.*

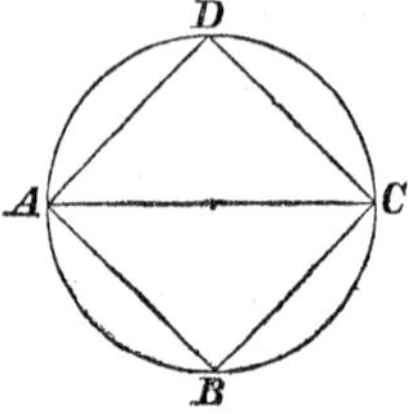


ANALYSIS. The diameter of the circle is the diagonal of the square *ABCD*, and is also the hypotenuse of the two right-angled triangles *ABC* and *CDA*. Hence,

*To find the side of a square inscribed in a given circle.*

**RULE.**

Square the diameter, divide the square by 2, and extract the square root of the quotient. *Or*,

Square the radius, multiply the square by 2, and extract the square root of the product.

NOTE 1. The pupil will readily perceive, upon an examination of this rule, that the area of a square inscribed in a circle is equal to ½ the square of the diameter, or twice the square of the radius of the circle.

**EXAMPLES FOR PRACTICE.**

1. The diameter of a circle is 10 inches; what is the side of its inscribed square? *Ans.* 7·07+ in.
2. What will be the side of a stick of square timber hewn from a log 2 feet in diameter?
3. The diameter of a circle is 3 yards; what is the area of its inscribed square? *Ans.* 4½ sq. yds.
4. The circumference of a circle is 11 inches; what is the side of its inscribed square?

NOTE 2. First find the diameter of the circle.

5. The side of a square is 4 feet; what is the diameter of its superscribing circle?
6. The area of a square is 49 inches; what is the radius of its superscribing circle? *Ans.* 4·94+ in.

---

**¶ 70.** Topic. Analysis. First rule. Second. Note 1.

**¶ 71.** *The diameter of a circle being given, to find the side of its inscribed octagon.*

Analysis. If from the radius *HO*, we subtract the line *KO*, (equal to ½ the side of the inscribed square,) the remainder *HK* will be the perpendicular, and the line *AK* or *GK*, (equal to ½ the side of the inscribed square,) will be the base of the right-angled triangle, *AKH* or *GKH*; and the hypotenuse *AH* or *GH*, will be one side of an octagon inscribed in the same circle. Hence,

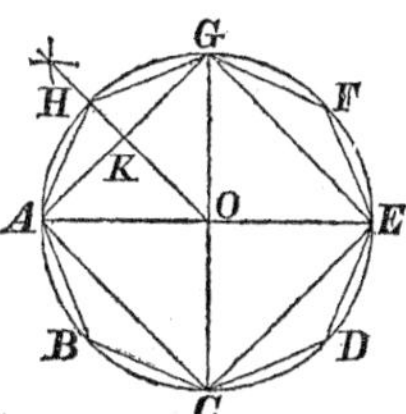


*To find the side of an octagon inscribed in a given circle.*

**RULE.**

I. Find the side of the inscribed square, by ¶ 70, rule, and subtract ½ of it from the radius of the circle.

II. Square this remainder, and also ½ the side of the inscribed square.

III. Add the squares together, and extract the square root of their sum.

Note. The side of a hexagon inscribed in a circle is equal to the radius of the circle.

**EXAMPLES FOR PRACTICE.**

1. The radius of a circle is 5 inches; what is the side of its inscribed octagon? *Ans.* 3·82+ in.

2. What will be the side of an octagon inscribed in a circle 24 inches in diameter?

3. A gentleman laid out a garden in the form of an octagon, the radius of whose circumscribing circle was 7 poles; what was the length of one side of the garden? *Ans.* 5 rds. 5 ft. 10·82+ in.

4. The circumference of a circle is 44 inches; what is the side of its inscribed hexagon?

5. The side of a hexagon inscribed in a circle is 7 inches; what is the side of a square, and also of an octagon, inscribed in the same circle?

**¶ 72.** *The two axes of an ellipse being given, to find the area.*

**RULE.**

Multiply the two diameters together, and their product by ·7854.

Note 1. This rule being deduced from the principles given in ¶63, a repetition of those principles in this place is deemed unnecessary.

---

¶ 71. Topic. Analysis. Rule. Note.
¶ 72. Topic. Analysis. Rule. Note 2.

**EXAMPLES FOR PRACTICE.**

1. What is the area of an ellipse whose axes are 15 and 22 feet?
2. The transverse diameter of an ellipse is 30 rods, and the conjugate diameter 20 rods; what is the area?
3. The area of an ellipse is 5497'8 feet, and the transverse diameter 100 feet; what is the conjugate diameter?

NOTE 2. When the area and one axis are given, the other axis may be found by dividing the area by '7854, and that quotient by the given axis. *Ans.* 70 ft.

4. The area of an ellipse is '7854 of a foot, and one axis is 9 inches; what is the other axis? *Ans.* $1\frac{1}{3}$ ft

**¶ 73.** *To find the diameter of a circle whose area shall be equal to the area of a given ellipse.*

**RULE.**

Multiply the axes of the given ellipse together, and extract the square root of the product.

NOTE. The same result will be obtained, by finding the area of the given ellipse, and from that area finding the diameter of the required circle.

For analysis of principles consult ¶¶ 63 and 64.

**EXAMPLES FOR PRACTICE.**

1. The axes of an ellipse are 35 and 48 feet; what is the diameter of a circle of equal area?
2. The axes of an ellipse are 20 and 30 yards; what is the diameter of a circle of equal area?
3. The area of an ellipse is 339'2928 square feet; what is the diameter of a circle of equal area? *Ans.* 20'784+ ft.

**¶ 74.** *To find the area of the space contained between the arcs of four equal circles, adjacent to each other, and all lying at an equal distance around a point.*

ANALYSIS. The square *ABCD* contains $\frac{1}{4}$ of the area of each circle, and also the space contained by the arcs *EF*, *FG*, *GH*, and *HE*. $\frac{1}{4}$ of the area of one circle multiplied by 4, is equal to one whole circle; and the area of the square, whose side is twice the radius of one circle, minus the area of one circle, is the area of the required space. Hence,

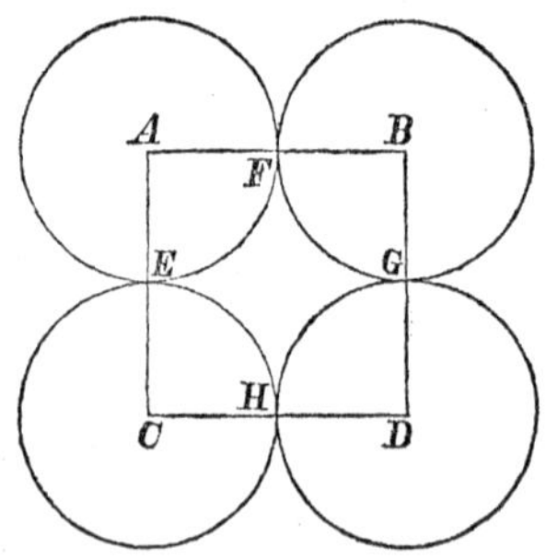


¶ 73. Topic. Analysis. Rule. Note.
¶ 74. Topic. Analysis. Rule.

*To find the area of the space contained between the arcs of 4 equal circles, adjacent to each other, and all lying at an equal distance around a point.*

**RULE.**

Subtract the area of one circle from the square of twice the radius.

**EXAMPLES FOR PRACTICE.**

1. The radius of each of 4 equal circles, lying at an equal distance from a point, is 9 inches; what is the area of the space contained between the arcs of the circles? *Ans.* 69·53+ sq. in.
2. The diameter of each of 4 equal circles, lying adjacent to each other, at an equal distance about a point, is 11 inches; what is the area of the space contained between the arcs of the circles?
3. What is the area of the space contained between the arcs of 4 equal circles, situated as in the last example, the area of one circle being 26·18 square chains? *Ans.* 7·153+ sq. C.

**¶ 75.** *To find the area of the space contained between the arcs of three equal circles, adjacent to each other, and all lying at an equal distance around a point.*

ANALYSIS. The equilateral triangle $ABC$ contains $\frac{1}{6}$ of the area of each circle, and also the area of the space contained between the arcs $DE$, $EF$, and $FD$. $\frac{1}{6}$ of the area of 3 equal circles is equal to $\frac{3}{6}$ or $\frac{1}{2}$ the area of one of them; and the area of the equilateral triangle, whose side is twice the radius of one circle, minus $\frac{1}{2}$ the area of one circle, is the area of the required space. Hence,

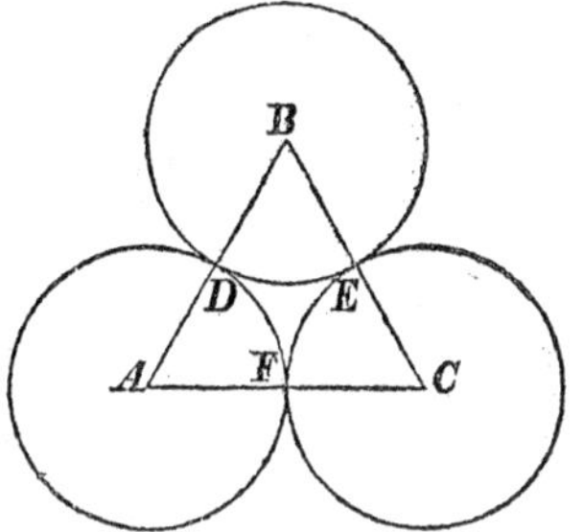


*To find the area of the space contained between the arcs of three equal circles, adjacent to each other, and all lying at an equal distance around a point.*

**RULE.**

Subtract $\frac{1}{2}$ the area of one circle from the area of an equilateral triangle whose side is twice the radius of one circle.

**EXAMPLES FOR PRACTICE.**

1. The radius of each of 3 adjacent circles, lying at an equal distance about a point, is $2\frac{1}{2}$ feet; what is the area of the space contained between their arcs? *Ans.* 1·00875 sq. ft.
2. The diameter of each of 3 circles, situated as in the last example, is $2\frac{2}{5}$ yards; what is the area of the space contained between their arcs?

---

¶ **75.** Topic. Analysis. Rule.

3. The circumference of each of three circles situated as in the last two examples, is 71 chains; what is the area of the space contained between their arcs? *Ans.* 20'566+ sq. C.

## ¶ 76. *To find the area of a circular ring.*

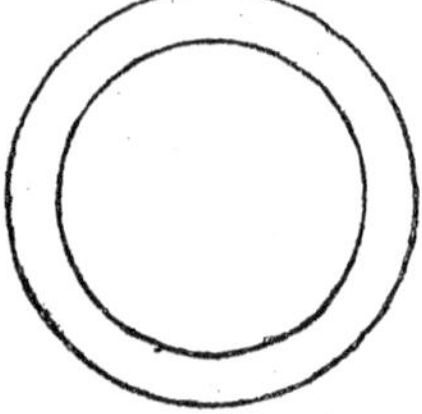

ANALYSIS. A circular ring is the space included between the circumferences of two concentric circles of different diameters.

Its area must evidently be equal to the area of the larger circle minus the area of the smaller. Hence,

*To find the area of a circular ring.*

### RULE.

Square the two diameters, subtract the less square from the greater, and multiply the remainder by '7854. *Or,*

Multiply the sum of the two diameters by their difference, and this product by '7854.

NOTE. The result in either case is the same as would be obtained by subtracting the area of the less circle from the area of the greater. The product of the sum of the diameters multiplied by their difference, is equal to the difference of their squares. (See ¶ 42.)

### EXAMPLES FOR PRACTICE.

1. Within a circular park 15 rods in diameter, is a circular pond 7 rods in diameter; what is the area of that part of the park not covered by the pond?

2. In a pleasure-ground is a circular pond, in the middle of which is a circular island; the diameter of the pond is 100 yards, and the circumference of the island the same; what is the area of the pond? *Ans.* 7058'22+ sq. yds.

3. A farmer has an elliptical orchard, whose axes are 300 and 200 yards, and he wishes to surround it with a wall 3 feet thick within the boundary line; how much land will be covered by the wall? *Ans.* 25 P. 26+ sq. yds.

## ¶ 77. *Similar curvilinear figures.*

The areas of similar curvilinear figures, as of similar rectilinear figures, are to each other as the squares of their similar radii, diameters, circumferences, curves, or linear dimensions; and the similar radii, diameters, circumferences, curves, or

---

¶ 76. Topic. Analysis. Rule. Note. Reasons for the operations directed in the note. Method of performing Ex. 3.

¶ 77. Topic. Principles. Where first demonstrated in this work. To what applied in that ¶.

linear dimensions are to each other as the square root of the quotient of the area of the greater figure divided by the area of the less. See ¶ 52.

### EXAMPLES FOR PRACTICE.

1. What must be the diameter of a circle, to contain 4 times the area of a circle 2 inches in diameter?

2. The circumference of a circle is 38 inches; what must be the circumference of a circle containing 16 times the area?

3. The diameter of a circle is 15 inches; what must be the diameter of that circle whose area is $\frac{1}{9}$ as great? *Ans.* 5 in.

4. The radius of one circle is 4 inches, and of another 9; what is the ratio of their areas? *Ans.* 16 to 81.

5. The radius of one sector is 2 feet, and of another similar one 6; what is the ratio of their arcs?

6. The arcs of two similar quadrants are 10 and 20 inches; what is the ratio of their radii, and of their areas?

7. The transverse diameter of an ellipse is 9 yards; what must be the transverse diameter of a similar ellipse, whose area is 25 times as great? *Ans.* 45 yds.

---

## PRACTICAL EXAMPLES

### In the Mensuration of Lines and Superficies.

**¶ 78.** 1. A mason plastered a room 22 ft. long, 19 ft. 9 in. wide, and 9 ft. 6 in. high; how much did he receive for the job, at $'18 a square yard? *Ans.* $24'55$\frac{1}{2}$.

2. How much must be paid for glazing 3 windows, each 3'5 ft. by 7'75 ft., at 10 cents a square foot? *Ans.* $8'13$\frac{3}{4}$.

3. A building 30 feet high stands on the bank of a stream 50 feet wide; what is the length of a ladder that will reach from the opposite bank of the stream to the top of the building? *Ans.* 58'3+ ft.

4. The base of a right-angled triangle is 62 feet, and the area 1240 feet; what is the length of the perpendicular? — of the hypotenuse?
*Ans.* Perpendicular 40 ft. Hypotenuse 66'66+ ft.

5. The perpendicular is 9, and the sum of the base and hypotenuse 19; what is the base? — the hypotenuse? *Ans.* { Base, $7\frac{7}{19}$. Hypotenuse, $11\frac{12}{19}$.

6. The base is 22 inches, and the sum of the perpendicular and hypotenuse 44 inches; what is the area? *Ans.* 181'5 sq. in.

7. The width of a certain barn is 24 feet, and the length of the rafters 13 feet; what will be the length of a purline beam extending from any rafter to the opposite one, and meeting the rafters at the distance of 7 feet from the bottom of each? *Ans.* $11\frac{1}{13}$ ft.

8. A field 60 rods long contains 15 acres; how many rods in length of the same field will be required for 9 acres? *Ans.* 36 rds.

9. I have a triangular board containing 94 square inches, the base, or longest side, being 23$\frac{1}{2}$ inches in length. I wish to divide this board into 3 parts, each having the same altitude as the whole triangle, and containing respectively 25, 33, and 36 square inches; what will be the length of the base of each piece? (See ¶ 47.) *Answers,* in order. 6'25 in.; 8'25 in.; 9 in.

10. The three sides of a triangular park measure respectively 20, 29, and 30 chains; how many acres in the park?
*Ans.* 27 A. 7 sq. C. 2 P. 268 sq. ft. 47'46+ sq. in.

11. What is the area of an equilateral triangle, whose side measures 44 inches? *Ans.* 838'3+ sq. in.

12. A certain rectangular field contains 82 A. 5 P. of land, and its length is to its breadth as 7 is to 3; what are the dimensions of the field?
*Ans.* 75 and 175 rods.

13. What length of a mahogany plank, 26 inches wide, will make 1½ square yards? *Ans.* 6'23+ ft.

14. A triangular field, whose legs measure 900 and 1775 links, rents for $37'50 per annum; how much is that an acre? *Ans.* $4'694+.

15. There is a house three stories high, with 7 windows in each story. Each window is 2 ft. 8 in. wide, and the hight of the windows in the first story is 6 ft. 10 in., in the second story 5 ft. 8 in., and in the third story 5 ft. 6 in.; what will the glazing come to, at $'14 a square foot?
*Ans.* $47'04,

16. What will the paving of a rectangular court-yard come to, at $'75 a square yard, the yard being 42 ft. 9 in. front, by 68 ft. 6 in. deep?
*Ans.* $244'03+.

17. The roof of a house is 52 ft. 8 in. long, and 45 ft. 9 in. from one eave to the other, across the ridge; what will the roofing cost, at $2'25 a square? *Ans.* $54'21+.

18. I have a stick of timber 3½ by 8½ inches, and I want another stick just twice as large, and 4½ inches thick; what must be its width?
*Ans.* 13'22+ in.

19. A wheelwright made a carriage-wheel 4 ft. 10 in. in diameter, and the rim consisted of 7 fellies; what was the length of each felly?
*Ans.* 2 ft. 2'03+ in.

20. The areas of 2 similar parallelograms are to each other as 9 to 7½, and the shorter side of the smaller parallelogram is 19 rods; what is the length of the shorter side of the larger parallelogram? *Ans.* 20'81+ rods.

21. The radius of a quadrant is 21 inches; what is the arc of another quadrant, whose area is ⅓ as great as the former? *Ans.* 19'04+ in.

22. What is the diameter of that circle whose area is 12 times as great as that of a circle 25 inches in diameter? *Ans.* 86'6+ in.

23. A pillar 7 inches in diameter is sufficient to sustain a certain weight; what must be the diameter of a pillar that shall sustain 10 times the weight, the length of the 2 pillars being the same? *Ans.* 22'13+ in.

24. Three pipes, each 3 inches bore, will fill a reservoir in a certain time; what must be the diameter of the bore of a pipe that will fill a reservoir 2½ times as large in the same time? *Ans.* 8'21+ in.

25. What is the diameter of a circular pond that covers ½ acre of surface? *Ans.* 10 rds. 1 ft. 6'3+ in.

26. What is the length of a cord, one end of which being fastened to a stake, and the other end to a horse's nose, will permit the horse to graze upon a semicircle containing just 1 acre of ground? *Ans.* 10 rds. 1'525+ ft.

27. There is a room 16 feet long, 15 feet wide, and 9 feet high; what is the nearest distance from any corner at the bottom to the diagonal corner at the top? *Ans.* 23'706+ ft.

28. A painter engaged to paint a church 86 feet long, 50 feet wide, 20 feet high to the top of the beams, and 17 feet from the beams to the ridge, for $'37½ per square yard; how much would the job come to, no deductions being made for windows, doors, &c., nor no additions for mouldings, cornices, &c.? *Ans.* $262'08+.

29. A portion of railroad 1 mile in length, passes through the farms of 3 men, as follows: 70 rods through the first farm, 115 rods through the second, and the remainder through the third. The owner of the first farm was awarded \$83'50 per acre as damages, the owner of the second farm \$92'37½ per acre, and the owner of the third farm \$100 per acre. Allowing the road to be 4 rods wide, how much did each man receive?

*Ans.* First man, \$146'12½; second, \$265'578+; third, \$337'50.

30. The shadow of a staff 3 feet long, at a certain hour of the day, measures 4 ft. 8 in.; what is the hight of that tree whose shadow at the same time measures 179 ft. 5 in.? *Ans.* 115 ft. 4$\frac{1}{14}$ in.

31. The wheels of a rail-car are each 2 ft. 8 in. in diameter; how many revolutions do they make in a minute, when the cars are running at the rate of 23 miles an hour? *Ans.* 241'599+.

32. The two diameters of a cylindroidal stick of timber are 5 and 9 inches; what is the mean diameter? *Ans.* 6'7+in.

33. What is the area of an elliptical field, whose diameters are 25 and 60 rods? *Ans.* 7 A. 3 sq. C. 10'1 P.

34. In a river are 6 circular islands, the diameters of 5 of which are 10'5, 16, 20'25, 26'75, and 32 rods, respectively. The area of the sixth is equal to the sum of the areas of all the others; what is its diameter? *Ans.* 50'158+rds.

35. A, B, C, and D, bought a grindstone 40 inches in diameter, for which they paid \$6'00, each paying an equal share. A first used the stone, till he had ground off his share; B then did the same, and so with C and D. What was the diameter of the stone when it came into the hands of B, C, and D, respectively?

*Answers*, in order. 34'64+in.; 28'28+in.; 20 in.

36. A gentleman laid out a circular pleasure-ground, which contained 28 P. 231 sq. ft. of land. He then laid out a graveled walk on the outer side of it, within the circle, which covered 4 sq. rds. of ground; what was the diameter of the pleasure-ground, and what the width of the walk? *Ans.* to last. 4'095+ft.

37. The mean distance of Mercury from the sun is 36000000 miles, of the Earth 95000000, and of Herschel 1827000000. If they all traveled in their orbits with the same velocity, how many revolutions would Mercury and the Earth each make round the Sun while Herschel made one? How many would Mercury make while the Earth made one?

38. Two horses are worked abreast upon the arm or sweep of a thrashing-machine. The sweep is 12 feet long, the horses are 3 feet apart, and the evener upon which they work is attached to the end of the sweep. Supposing the horses to make 2 circuits in a minute, and to work 8 full hours a day, how much farther will the horse working upon the outside travel, during the day, than the one working upon the inside?

*Ans.* 3 mi. 126 rds. 11'5+ft.

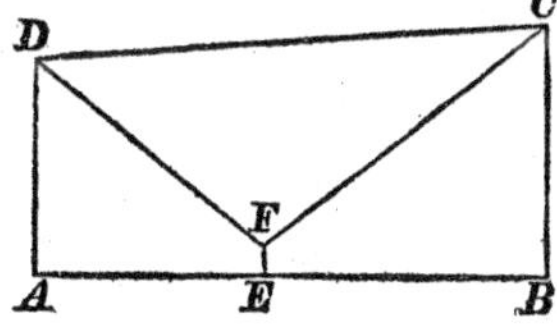


39. *AD* and *BC* are the fronts of two houses, standing on opposite sides of a public square; and *EF* is a post standing in the square, in a right line between the houses. The hight of *AD* is 55 feet, and of *BC* 64 feet. The distance from the foot of the post to the base of the house *BC*, is 76 feet; from the top of the post to the top of the house *BC*, 95

feet; and from the top of the post to the top of the house *AD*, 80 feet. What is the hight of the post? — the distance from the base of one house to the base of the other? — from the top of one house to the top of the other? *Answers*, in order. 7 ft.; 140 ft.; 140'28+ ft.

---

# MENSURATION OF SOLIDS.

**¶ 79.** *To find the cubic contents of a prism, cube, parallelopiped, cylinder, or cylindroid.*

**RULE.**

Multiply the area of one end by the length, or the area of the base by the altitude.

NOTE 1. For analysis of principles, see Revised Arithmetic, ¶ 51.

**EXAMPLES FOR PRACTICE.**

1. The side of a cubic block measures 8 inches; how many cubic inches does it contain?
2. How many cubic feet in a cube whose side measures 11 feet?
3. The end of a square prism is 10 inches square, and the length is $2\frac{1}{2}$ feet = 30 inches; how many cubic feet does it contain?

NOTE 2. When the three dimensions are in inches, divide the cubic contents by 1728; when two dimensions are in inches and the third in feet, divide by 144; and when two dimensions are in feet and the third in inches, divide by 12; and in either case the quotient will be cubic feet. Why?

4. The end of a prism 20 feet long is a right-angled triangle, the two shorter sides of which measure 9 and 12 inches; what are the cubic contents of the prism?
5. What are the contents of a parallelopiped 15 feet long, 3 feet wide, and 11 inches thick? *Ans.* $41\frac{1}{4}$ cu. ft.
6. What is the solidity of a cylinder 7 feet long, and 2 feet in diameter?
7. What are the contents of a log 31 feet long, and $17\frac{1}{2}$ inches in diameter? *Ans.* 51'78+ cu. ft.
8. What are the solid contents of a stick of timber 28 feet long, and 8 inches square? *Ans.* 12 cu. ft. 768 cu. in.
9. A stick of timber is 25 ft. 3 in. long, 1 ft. 8 in. wide, and 18 in. thick; how much will it come to, at 8 cents per cubic foot?

NOTE 3. Consult ¶ 57, Note 4. *Ans.* $5'05.

10. What is the solidity of a block of marble 10 ft. long, $5\frac{3}{4}$ ft. wide, and $3\frac{1}{8}$ ft. thick?
11. A cistern is $5\frac{1}{2}$ feet in diameter, and 8 feet deep; how many standard gallons will it contain? *Ans.* 1421'7984 gal.
12. The diameters of a cylindroidal tube 200 feet long, are 3 and 5 inches; how many standard gallons will it contain? *Ans.* 122'4 gals.
13. The side of the base of a regular hexagonal prism is 9 inches, and the altitude is 14 feet; what is the solidity?

---

**¶ 79.** Topic. Rule. Analysis. Note 2. Note 4.

14. What is the solidity of a regular octagonal pillar 26 feet long, one side of which measures 7 inches? (See ¶ 54.) *Ans.* 42·718+ cu. ft.

NOTE 4. The superficial contents of any of the figures named in this ¶ may be obtained, by multiplying the circumference or the girth of one end of the figure by the length, and to the product adding the areas of the two ends. Why?

15. What is the surface of a cube whose side is 4 feet?

16. The end of a prism 25 feet long is an equilateral triangle, the side of which measures 16 inches; what is the area of the prism? *Ans.* 101·539+ sq. ft.

17. What is the surface of a prism 18 feet long, and 21 inches square?

18. What are the superficial contents of a round pillar 14 inches in diameter, and 30 feet long? *Ans.* 102 sq. ft. 20 sq. in.

## ¶ 80. *To find the cubic contents of a pyramid or a cone.*

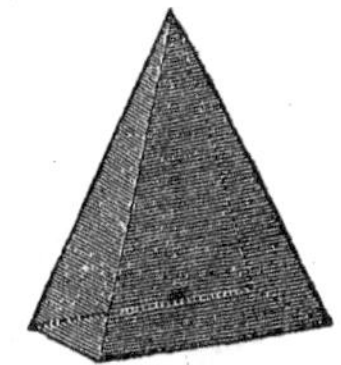

ANALYSIS. The cubic contents of any pyramid, of a given base and altitude, are equal to ⅓ of the cubic contents of a prism having the same base and altitude. And,

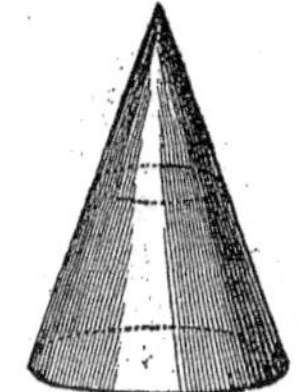

The cubic contents of any cone, of a given base and altitude, are equal to ⅓ of the cubic contents of a cylinder having the same base and altitude. Hence,

### *To find the cubic contents of a pyramid or a cone.*

**RULE.**

Multiply the area of the base by ⅓ of the altitude; *or*,

Multiply the area of the base by the altitude, and take ⅓ of the product.

NOTE 1. The correctness of this rule may be verified by rule ¶ 84.

**EXAMPLES FOR PRACTICE.**

1. What is the solidity of a pyramid 15 feet square at the base, and 40 feet high?

2. Each side of the base of a triangular pyramid is 30 inches, and the altitude is 4 feet; what are the cubic contents? *Ans.* 10·825 cu. ft.

---

¶ 80. Topic. Analysis. Rule. Proof. Note 2. Note 3. Note 4.

3. The area of the base of an octagonal pyramid is 78 square feet, and the altitude is $19\frac{1}{2}$ feet; what is the solidity?

4. The base of a cone is 7 feet in diameter, and the altitude is 16 feet 9 inches; what are the solid contents?

5. The altitude of a cone is 5 feet, and the circumference of the base $5\frac{1}{12}$ feet; what are the cubic contents? *Ans.* 4'64295+ cu. ft.

6. The slant hight of a cone is 18 inches, and the diameter of the base 15 inches; what is the solidity?

NOTE 2. The slant hight of a cone is the distance from the vertex to the circumference of the base, and the slant hight of a pyramid is the distance from the vertex to the middle of one side of the base.

7. What is the solidity of a pyramid 30 feet square at the base, the slant hight being 25 feet? *Ans.* 6000 cu. ft.

NOTE 3. The outside of a pyramid and a cone is called the lateral or convex surface, the area of which may be found by multiplying ½ the circumference or girth of the base by the slant hight; and, when the entire surface is required, to the product adding the area of the base. Why?

8. The slant hight of a pyramid is 11 inches, and the base is 4 inches square; what is the entire surface?

9. What is the area of a triangular pyramid, each side of the base measuring 30 feet, and the slant hight 42 feet? *Ans.* 2279'7114+ sq. ft.

10. What is the lateral surface of a cone, the slant hight being 38 inches, and the circumference of the base 40 inches? *Ans.* 5 sq. ft. 40 sq. in.

11. The solidity of a cone is 214'87235 cubic feet, and the altitude 16'75 feet; what is the diameter of the base?

NOTE 4. This example involves the principles of the rule, but they must be applied in a reverse order.

12. The cubic contents of a square pyramid are 3000 cubic inches, and the altitude is 40 inches; what is the length of one side of the base?

13. The area of the base of a hexagonal pyramid is 259'8076 square yards, and the solidity 1299'038 cubic yards; what is the altitude? *Ans.* 45 ft.

**¶ 81.** *To find the hight of a pyramid or cone, of which a given frustrum is a part.*

### RULE.

I. *For the pyramid;* — Say, the difference between one side of the top and one side of the base : one side of the base : : the altitude of the frustrum : the altitude of the pyramid. *Or,*

The difference between the perimeter of the top and the perimeter of the base : the perimeter of the base : : the altitude of the frustrum : the altitude of the pyramid.

II. *For the cone;* — Say, the difference between the diameters (or the radii) of the top and base : the diameter (or the

---

¶ 81. Topic. Analysis. Rule.

radius) of the base : : the altitude of the frustrum : the altitude of the cone. *Or*,

The difference between the peripheries of the top and base : the periphery of the base : : the altitude of the frustrum : the altitude of the cone.

NOTE. For analysis of principles, see ¶ 44.

### EXAMPLES FOR PRACTICE.

1. The base of the frustrum of a pyramid is 8 feet square, the top 3 feet square, and the altitude 15 feet; what was the hight of the pyramid? *Ans.* 24 ft.

2. The base of the frustrum of a hexagonal pyramid is 22 inches on each side, the top 9 inches on each side, and the altitude 5 feet; what was the altitude of the pyramid?

3. The perimeter of the base of a decagonal frustrum is 8 feet 4 inches, the perimeter of the top 2 feet 1 inch, and the altitude 10 feet; what was the altitude of the pyramid?

4. The diameter of the base of the frustrum of a cone is 5 feet, the diameter of the top 4 feet, and the altitude 20 feet; what was the altitude of the cone?

5. The radius of the base of the frustrum of a cone is 17 inches, the radius of the top 14 inches, and the altitude 5 inches; what was the altitude of the cone?

6. The circumference of the base of the frustrum of a cone is 47 feet, the circumference of the top 41 feet, and the altitude $28\frac{2}{3}$ feet; what was the altitude of the cone? *Ans.* 224 ft. $6\frac{2}{3}$ in.

**¶ 82.** *To find the solidity of the frustrum of a pyramid or cone.*

### RULE.

I. Find the hight of the pyramid or cone of which the given frustrum is a part, by ¶ 81.

II. Find the cubic contents of the pyramid or cone, and also of the segment, by ¶ 80.

III. Subtract the cubic contents of the segment from the cubic contents of the entire pyramid; the remainder will be the cubic contents of the frustrum.

NOTE 1. The pupil will readily comprehend the reasons for each step in the rule.

### EXAMPLES FOR PRACTICE.

1. What is the solidity of the frustrum of a square pyramid, one side of the greater end being 18 inches, one side of the smaller end 15 inches, and the altitude 5 feet? *Ans.* $9\frac{23}{48}$ cu. ft.

2. What is the solidity of the frustrum of a hexagonal pyramid, the side of the greater end being 3 feet, that of the smaller end 2 feet, and the altitude 12 feet? (See ¶ 54.)

---

¶ **82.** Topic. Analysis. Rule. Note 2.

3. What is the solidity of the frustrum of a cone, the diameter of the greater end being 4 feet, that of the smaller end $2\frac{1}{2}$ feet, and the altitude 11 feet 3 inches? *Ans.* 94'9843+ cu. ft.

4. What is the solidity of the frustrum of a cone, the circumference of the greater end being 83 inches, that of the smaller end 54 inches, and the altitude 12 feet?

5. A man has a vessel in the form of the frustrum of a square pyramid; the lower end is 30 inches square, the upper end 20 inches square, and the altitude 4 feet; how many dry gallons will it contain? *Ans.* $113\frac{1}{3}$ dry gal.

6. The diameter of the top of a tub, in the form of an inverted frustrum of a cone, is 40 inches, the diameter of the bottom 30 inches, and the altitude 5 feet; what are the contents in standard gallons? *Ans.* 251'6 gal.

NOTE 2. The lateral surface of the frustrum of a pyramid or a cone may be found, by multiplying $\frac{1}{2}$ the sum of the girths of the two ends by the slant hight, and when the entire surface is required, to the product adding the areas of the two ends. Why?

7. What is the area of the frustrum of a pyramid whose slant hight is 8 feet, the base 4 feet square, and the top 2 feet 3 inches square? *Ans.* 121 sq. ft. 9 sq. in.

8. What is the area of the frustrum of a triangular pyramid, whose slant hight is 16 inches, each side of the base 3 feet, and each side of the top $1\frac{1}{2}$ feet?

9. What is the convex surface of the frustrum of a cone, whose slant hight is 18 inches, the circumference of the base 38 inches, and of the top 28 inches? *Ans.* 4 sq. ft. 18 sq. in.

## ¶ 83. *To find the superficial and the cubic contents of the regular solids.*

ANALYSIS. Each of the regular solids may be divided into as many equal pyramids as the solid has faces, the base of each pyramid being a face of the solid, and the altitude the perpendicular distance from the centre of one face to the centre of the solid.

Since it is somewhat difficult to find the altitude of the pyramids of which each regular solid is composed, the following table has been prepared, by the aid of which the superficies, and the solidity of any regular solid, may readily be found, by having one side and the number of sides given.

| Length of one side. | Number of sides. | Names of Solids. | Superficies. | Solidity. |
|---|---|---|---|---|
| 1 | 4 | Tetraedron, | 1'732051 | '117851 |
| 1 | 6 | Hexaedron, | 6'000000 | 1'000000 |
| 1 | 8 | Octaedron, | 3'464102 | '471404 |
| 1 | 12 | Dodecaedron, | 20'645729 | 7'663119 |
| 1 | 20 | Icosaedron, | 8'660254 | 2'181695 |

It has been shown, ¶¶ 52 and 77, that the areas of similar figures are to each other as the squares of their similar dimensions. It is also true

¶ **83.** Topic. Analysis. Reasons for preparing a table. Its construction. Similar solids. Rule for use of table.

that the solidities of similar bodies are to each other as the cubes of their similar dimensions.* Hence,

*To find the superficies or the solidity of any regular solid, by the table.*

**RULE.**

I. *For the superficies;* — Multiply the square of one side by the tabular number of the superficies of a similar solid.

II. *For the solidity;* — Multiply the cube of one side by the tabular number of the solidity of a similar solid.

**EXAMPLES FOR PRACTICE.**

1. The side of a tetraedron measures 7 inches; what are its superficies and solidity?
2. The side of an octaedron measures 4 inches; what are its superficies and solidity?
3. What are the superficies and the solidity of a dodecaedron, one side of which measures $4\frac{1}{2}$ feet?
4. What are the superficies and the solidity of an icosaedron, one side of which measures 3 inches?

**¶ 84.** *To find the cubic contents of any irregular solid.*

**RULE.**

I. Place the solid in a tub, cylinder, cubical box, or any other vessel whose contents can be ascertained, and then fill the vessel with water.

II. From the cubic contents of the vessel subtract the cubic contents of the water put in to fill the vessel; the remainder will be the cubic contents required.

NOTE. Any vessel may be filled with water, and the body whose contents are required may then be immersed in the water; the quantity of water which the body displaces, or which runs over the sides of the vessel, will be equal in bulk or cubic contents to the figure immersed.

**EXAMPLES FOR PRACTICE.**

1. A blacksmith's anvil was put into a tub, the capacity of which was $8\frac{1}{4}$ wine gallons, and the tub was afterwards filled with 6 gal. 3 qts. 1 pt. of water; what was the solidity of the anvil? *Ans.* $317\frac{5}{8}$ cu. in.
2. A chain was put into a cubical box whose inside measured 8 inches, and the box was afterwards filled with $3\frac{3}{4}$ quarts of water; what were the cubic contents of the chain? *Ans.* $395\frac{7}{16}$ cu. in.
3. A pig of iron was put into a cylinder 3 feet long, and 7 inches in

---

¶ **84.** Topic. Analysis. Rule. Method explained in note.

* It may be well to test the pupil's comprehension of this truth, by giving him a few practical examples in Similar Solids. See ¶¶ 52 and 77.

diameter, and the cylinder was then filled with 4 gallons of water; what were the cubic contents of the iron? *Ans.* 461'4456 cu. in.

## ¶ 85. *To find the area of a sphere or globe.*

It is demonstrated in Geometry, that the area of a sphere or globe is 4 times as great as the area of a circle of the same diameter. Hence,

*To find the area of a sphere.*

**RULE.**

Multiply 4 times the square of the diameter by '7854. *Or,* Multiply the whole circumference by the whole diameter.

NOTE. A knowledge of Geometry is necessary to a full understanding of the principles of this rule.

**EXAMPLES FOR PRACTICE.**

1. How many square inches on the surface of a globe 15 inches in diameter?
2. How many square feet on the surface of a sphere 4 feet in diameter?
3. The diameter of the earth is 7911 miles; what is the area, rejecting fractions of a mile in the circumference? *Ans.* 196612083 sq. mi.

## ¶ 86. *To find the solidity of a sphere.*

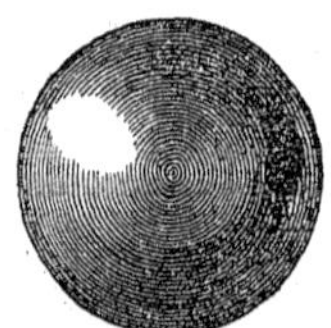

ANALYSIS. Any sphere may be supposed to be divided into an infinite number of pyramids, whose vertices all meet in the centre of the sphere, and the areas of whose bases form the area of the sphere. Since the solidity of any pyramid is equal to the area of its base multiplied by $\frac{1}{3}$ of its altitude, (¶ 80,) the solidity of all the pyramids into which any sphere may be supposed to be divided, is equal to the areas of all their bases (which is the surface of the sphere) multiplied by $\frac{1}{3}$ of the altitude of one of them, or $\frac{1}{6}$ of the diameter of the sphere. Hence,

*To find the solidity of a sphere.*

**RULE.**

Multiply the area of the sphere by $\frac{1}{6}$ of the diameter.

NOTE 1. Any sphere is equal to $\frac{2}{3}$ of a cylinder whose diameter and altitude are each equal to the diameter of the given sphere. Hence, the solidity of a

---

¶ 85. Topic. Analysis. Rule. Note.
¶ 86. Topic. Analysis. Rule. Note 1.

sphere may be found, by taking ⅔ of the solidity of a cylinder whose diameter and altitude are each equal to the diameter of the sphere; *i. e.*, by multiplying the cube of the diameter of the sphere by '7854, and taking ⅔ of the product.

### EXAMPLES FOR PRACTICE.

1. The diameter of a sphere is 18 inches; what is its solidity?
2. What is the solidity of an ivory ball 2 inches in diameter?
3. What is the solidity of a wicket ball 18½ inches in circumference?
4. The diameter of the earth is 7911 miles; what is its solidity, rejecting fractions of a mile in the circumference? *Ans.* 259233031435½ cu. mi.
5. What is the solidity of a ball that can just be put into a cylindrical cup 5 inches in diameter, and 5 inches deep?
6. What is the solidity of a hemisphere 12 inches in diameter? *Ans.* 358'14+ cu. in.
7. The solidity of a sphere is 65'45 cu. in.; what is its diameter?

NOTE 2. This example may be performed by the principles given in Note 1; but they must be applied in a reverse order.

---

## GAUGING.

**¶ 87.** *Gauging* is measuring the capacity of barrels, casks, hogsheads, &c.

ANALYSIS. The mean diameter of a barrel, cask, &c., may be found, by adding to the head diameter ⅔, or, if the staves be but little curving, $\frac{6}{10}$ of the difference between the head and bung diameters. The cask will then be reduced to a cylinder. Now, if the square of the mean diameter expressed in inches be multiplied by '7854, the product will be the area of one end, ¶63, and this area multiplied by the length in inches, will give the cubic contents in cubic inches. Hence,

### *To gauge or measure a cask.*

### RULE.

I. *For the capacity in cubic inches;*—Multiply the area of the mean diameter in inches by the length in inches.

II. *For the capacity in standard or wine gallons;*—Divide the capacity in cubic inches by 231. ¶ 20.

III. *For the capacity in bushels;*—Divide the capacity in cubic inches by 2150'4. ¶ 24.

NOTE. The capacity in dry gallons may be found, by dividing the capacity in cubic inches by 268'8, ¶ 24; and in beer gallons, by dividing by 282, ¶ 22.

### EXAMPLES FOR PRACTICE.

1. The head diameter of a cask is 22 inches, the bung diameter

---

**¶ 87.** Topic. Gauging. Analysis. Rule. Note.

28 inches, and the length 31 inches; how many standard or wine gallons will it contain? How many bushels? How many beer gallons?

*Ans.* to first two. 71·2504 wine gal.; 7·6538+ bush.

2. The head diameter of a cask is 30 inches, the bung diameter 35 inches, and the length 40 inches; what is its capacity in standard liquid gallons? — in bushels?

---

## Timber Measure.

**¶ 88.** Square or hewn timber is sometimes bought and sold by the cubic foot, and is sometimes reduced to standard board measure.

*To find the number of cubic feet in any stick of hewn timber which does not taper.*

For rule and principles, consult ¶ 79.

*To find the number of feet, board measure, in any stick of hewn timber which does not taper.*

For rule and principles, see ¶¶ 79 and 37.

### EXAMPLES FOR PRACTICE.

1. How many cubic feet in a stick of timber 50 feet long, and 7 by 10 inches?

2. How many cubic feet in a stick of timber 40 feet long, and 22 by 27 inches?

3. How many feet, board measure, in a stick of timber 60 feet long, and 8 by 14 inches?

4. How many feet, board measure, in a stick of timber 35 feet long, 15 inches wide, and 12 inches thick?

**¶ 89.** *To find the contents of a four-sided stick of timber, which tapers upon two opposite sides only.*

Analysis. If a stick of timber tapering upon two opposite sides only, be sawn into boards in a line perpendicular to the tapering sides, it will make a certain number of boards of uniform length, and all tapering alike. Hence,

*To find the contents of a four-sided stick of timber which tapers upon the opposite sides only.*

### RULE.

I. Divide the sum of the widths of the two ends of either

---

**¶ 88.** Measure by which square timber is bought and sold. First rule. Analysis. Second rule. Analysis.

**¶ 89.** Topic. Analysis. Rule.

tapering side by 2; the quotient will be the mean width of the tapering side.

II. Proceed in all other respects as directed in ¶ 79.

### EXAMPLES FOR PRACTICE.

1. A stick of timber is 24 feet long, 15 inches thick, 12 inches wide at one end, and 8 at the other; what are its cubic contents? What its contents in board measure?

2. A stick of timber is 18 feet long, 12 inches wide, 15 inches thick at one end, and 10 at the other; what are its cubic contents? What its contents in board measure?

## ¶ 90. *To find the contents of a stick of timber which tapers uniformly upon all sides.*

ANALYSIS. A stick of timber tapering uniformly upon all sides, is either a pyramid or the frustrum of a pyramid, and consequently must be measured by the same principles. See ¶¶ 80 and 82.

### EXAMPLES FOR PRACTICE.

1. A stick of hewn timber is 31·5 feet long, 18 inches square at one end, and 8 inches square at the other; what are its cubic contents?

2. A stick of timber 20 feet long is 12 inches square at one end, 15 inches square at the other, and 20 inches square in the middle; what are its contents in board measure? *Ans.* 474 sq. ft. 104 sq. in.

## ¶ 91. *To find the number of cubic feet of timber any log will make when hewn square.*

### RULE.

Find the area of the inscribed square of the smaller end of the log, and multiply this area by the length.

NOTE 1. For explanation of principles, see ¶ ¶ 70 and 70.

NOTE 2. The cubic contents of a log may be found by the rule for finding the cubic contents of a cylinder, ¶ 79.

NOTE 3. In measuring a log for any purpose, it is considered as a cylinder of the same diameter as the smaller end of the log. Hence, the diameter of the smaller end of the log must always be taken; and, if the end be elliptical, the shorter diameter must be taken, not the longer one.

### EXAMPLES FOR PRACTICE.

1. A log is 28 feet long, and 20 inches in diameter; how many feet of timber will it make when hewn square?

---

**¶ 90.** Topic. Analysis. Principles referred to. Rules for the operations.
**¶ 91.** Topic. Analysis. Rule. Note 2. Note 3.

2. A log 20 feet long, and 10 inches in diameter, was hewn square; how many cubic feet were cut away? *Ans.* 3 cu. ft. 1665'6 cu. in.

3. A stick of timber 12 feet long, and 14 inches in diameter, was hewn into a hexagonal form; how many cubic feet did the hexagon contain? *Ans.* 10'60881+ cu. ft.

**¶ 92.** *To find the number of feet of boards that can be sawn from any log of a given diameter.*

Logs for sawing are measured in three different ways, which we will designate as 1st, 2d, and 3d methods.

## First Method.

ANALYSIS. The allowances to be made in this method are,

1st. *For slabs.* This is an allowance of 2 inches on each side, or 4 inches of the diameter of any log not exceeding 2 feet in diameter; and 3 inches on each side, or 6 inches of the diameter of any log more than 2 feet in diameter.

2nd. *For saw space.* This is an allowance of $\frac{1}{4}$ of an inch for each time the saw passes through the log. In sawing boards of the standard thickness, the saw cuts away just $\frac{1}{5}$ of the log after the slabs are removed.

3d. *For wane.* This is an allowance of 1 board for any log not exceeding 2 feet in diameter; and of 2 boards for any log more than 2 feet in diameter. This allowance is made to cover the loss that would otherwise arise, from estimating the wane-edged boards the same as those that are square-edged.

EXAMPLE. How many feet of boards can be cut from a log 12 feet long, and 2 feet = 24 inches in diameter?

SOLUTION. 24 in. (diameter) — 4 in. (slabs) = 20 in. for sawing. 20 in. — 4 in. ($\frac{1}{5}$ of 20 in. for saw space) = 16 in. thickness of all the boards, which — 1 (for wane) = 15 boards that measure. We now have the log reduced to a stock of 15 boards, each 12 feet long, and 20 inches wide. 20 in. (width of 1 board) × 15 (no. of boards) = 300 in. = 25 ft., the whole width of the boards. Then, 25 ft. (width) × 12 ft. (length) = 300 sq. ft., the *Ans.* Hence,

*For the first method.*

### RULE.

I. Make the customary allowances for slabs, saw space, and wane; the remainder is the number of standard boards that can

---

¶ **92.** Topic. Methods of measuring logs for sawing. Analysis of first method. 1st allowance. 2d. 3d. Solution of example. Rule.

be sawn from the log, the width of each being the diameter of the log minus the allowance for slabs.

II. Find the contents of the stock of boards to which the log is reduced, as directed in ¶ 57.

## ¶ 93. Second Method.

ANALYSIS. In this method no deductions or allowances are made, but the log is reduced to a stick of square timber, by ¶ 91, and then to board measure. It is estimated that in reducing the log to a stick of square timber, the siding or wane-edged boards will equal in quantity the loss by saw space. In this estimate there is a little advantage in favor of the buyer; but when we consider that some logs are crooked, some rotten, some hollow, &c., it seems but just to have the advantage in his favor. Hence,

*For the second method.*

RULE.

Find the number of feet, board measure, that the log will make when hewn square.

NOTE. This method is the one generally adopted in the lumbering districts of New York and Pennsylvania. Some of the most accurate business men in the lumber trade, after trying various methods, say, that this method is as nearly correct for all logs over 13 inches in diameter, as any other method that has been presented.

## ¶ 94. Third Method.

ANALYSIS. In this method a slab of 1 inch in thickness is first taken off from one side, and the log is then sawn up into wane-edged boards. The slab upon the opposite side to the one first taken off, is never allowed to be less than $\frac{3}{4}$ of an inch, nor over 2 inches, in thickness. From an actual measurement of stocks of boards sawn from logs varying from 7 to 36 inches in diameter, by this method, it has been found that

1st. The width of the second board is the average width of all the boards sawn from any log from 7 to 12 inches in diameter.

2nd. The width of the third board is the average width of all the boards sawn from any log from 12 to 24 inches in diameter. And

3rd. The width of the fourth board is the average width of all the boards sawn from any log from 24 to 36 inches in diameter. Hence,

---

¶ 93. Analysis of second method. Rule. Note.

¶ 94. Analysis of third method. Stocks of boards sawn from logs from 7 to 12 inches in diameter, by this method. From 12 to 24 inches in diameter. From 24 to 36 inches. Rule. Note 1. Note 2.

*For the third method.*

### RULE.

I. Multiply the width of the average board, in inches, by the number of boards.

II. Multiply this product by the length of the log or stock of boards in feet, and divide the product by 12.

NOTE 1. Consult ¶ 58.

NOTE 2. The manner of sawing logs described in this ¶ is in quite extensive practice in New England.

### EXAMPLES FOR PRACTICE.

1. The diameter of a log is 31 inches, and the length 13 feet; how many feet of boards will it make, estimating by the first method? By the second? *Ans.* By 1st method, $487\frac{1}{2}$ sq. ft.; by 2d, $520\frac{13}{24}$ sq. ft.

2. The diameter of a log is 40 inches, and the length 10 feet; how many feet of boards will it make, estimating by the first method? By the second? *Answers,* in order. $708\frac{1}{3}$ sq. ft.; $666\frac{2}{3}$ sq. ft.

3. How many feet of boards may be sawn from a log 19 inches in diameter, and 16 feet long, estimating by the first method? By the 2d? *Answers,* in order. 220 sq. ft.; $240\frac{2}{3}$ sq. ft.

## ¶ 95. *To find how many bushels of grain may be put into a bin of a given size.*

### RULE.

Divide the cubic contents of the bin in inches by 2150'4.

NOTE 1. If the pupil does not readily recognize the principles involved in the rule, he should review ¶¶ 24 and 79.

NOTE 2. 2150'4 cu. in. (= 1 bush.) — 430'08 cu. in. ($=\frac{1}{5}$ of a bush.) = 1720'32 cu. in., or 7'68 cu. in. less than one cubic foot. Hence, if the cubic feet in any bin be diminished $\frac{1}{5}$, the remainder will be the number of bushels which the bin will contain. This result, although not strictly correct, is sufficiently accurate for all practical purposes.

### EXAMPLES FOR PRACTICE.

1. I have a bin 5 feet long, 4 feet 8 inches wide, and 2 feet 10 inches deep; how many bushels of grain will it hold?

2. A farmer has a bin 8 feet long, 4 feet wide, and 5 feet deep; how many bushels of grain will it hold?

3. A bin 5 feet long, and 4 feet 8 inches wide, contains $53\frac{1}{3}$ bushels of grain; what is its depth?

4. A farmer wishes to construct a bin 8 feet long, and 5 feet deep, and he will have it hold 128 bushels; what must be its width?

---

¶ 95. Topic. Analysis. Rule. Note 2. Note 3.

NOTE 3. 1728 cu. in. (= 1 cu. ft.) + 432 cu. in. (= ¼ of a cu. ft.) = 2160 cu. in., or 9'6 cu. in. more than 1 bushel. Hence, if the number of bushels in any bin be increased ¼, the sum will be the contents in cubic feet. This result, although not strictly correct, is sufficiently accurate for all practical purposes.

## ¶ 96. TABLE,

*Showing the inside dimensions of any box of a given capacity, in dry measure.*

| Length. | Width. | Depth. | Capacity. |
|---|---|---|---|
| 4 inches. | 4 inches. | 4'2 inches. | 1 quart. |
| 7 " | 4 " | 4'8 " | 2 quarts = ½ gallon. |
| 8 " | 8 " | 4'2 " | 1 gallon = ½ peck. |
| 8'4 " | 8 " | 8 " | 1 peck. |
| 12 " | 11'2 " | 8 " | 2 pecks = ½ bushel. |
| 16'8 " | 16 " | 8 " | 1 bushel. |

## ¶ 97. *To find the side of the greatest cube that can be cut from any sphere.*

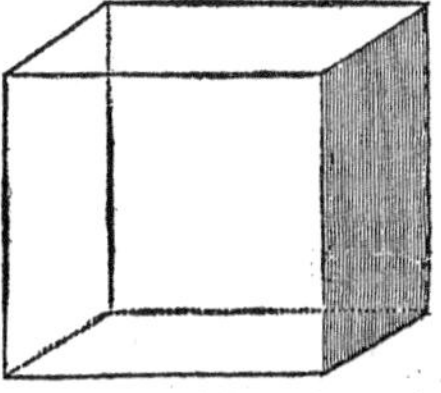

ANALYSIS. The diameter or axis of the sphere is the length of a diagonal running from any lower corner of the cube to the diagonal corner at the top. Let us first see how this diagonal of any cube is obtained. We first square a side of the cube, double it, and then extract the square root. This gives us the diagonal of one face of the cube. We next square this diagonal, (which square will contain twice the square of one side,) add to it the square of one side of the cube, (making the sum 3 times the square of one side of the cube,) and then extract the square root of this sum. Hence,

*To find the side of the greatest cube that can be cut from any sphere.*

### RULE.

Divide the square of the diameter of the sphere by 3, and extract the square root of the quotient.

### EXAMPLES FOR PRACTICE.

1. What is the size of the largest cube that can be cut from a sphere 15 inches in diameter? *Ans.* 8'66+ in.

2. I have a globe 25 inches in diameter, and I wish to cut from it the largest possible cube; how much of the sphere will I cut away? *Ans.* 5794'98+ cu. in.

---

¶ 96. Table.
¶ 97. Topic. Analysis. Rule.

**¶ 98.** *To find the weight of a lead or of a cast-iron ball.*

ANALYSIS. The cubic contents, and consequently the weights of similar bodies of the same substance, are to each other as the cubes of their similar dimensions. (¶¶ 52, 77, and 83.) A leaden ball 1 inch in diameter weighs $\frac{3}{14}$ of a pound; and a cast-iron ball 4 inches in diameter weighs 9 pounds. Hence,

*To find the weight of a lead or of a cast-iron ball.*

RULE.

I. *For a leaden ball;* — $1^3$ : the cube of the given diameter : : $\frac{3}{14}$ lb. : the weight of the given ball.

II. *For a cast-iron ball;* — $4^3$ : the cube of the given diameter : : 9 lbs. : the weight of the given ball.

EXAMPLES FOR PRACTICE.

1. What is the weight of a leaden ball 5 inches in diameter?
*Ans.* $26\frac{11}{14}$ lbs.
2. What is the weight of a cast-iron ball 7 inches in diameter?
*Ans.* $48\frac{15}{64}$ lbs.
3. A cast-iron ball weighs 22·5 pounds; what is its diameter?
*Ans.* 5·42+ in.

## PRACTICAL EXAMPLES

### IN THE MENSURATION OF SOLIDS.

**¶ 99.** 1. How many 5 inch cubes are equal to a 20 inch cube?
*Ans.* 64.

2. A stick of timber is 30 feet long, 10 inches wide, 10 inches thick at one end, and 7 at the other; what are its cubic contents?
*Ans.* 17 cu. ft. 102 cu. in.

3. A stick of timber is 23 feet long, 20 inches thick, 14 inches wide at each end, and 20 in the middle; what are its contents in board measure?
*Ans.* $651\frac{2}{3}$ sq. ft.

4. I have 4 sticks of timber of the same length, each 25 inches wide, and whose thicknesses are 21, 10, 6, and 3·5 inches, respectively. I wish a square stick of the same length, and containing as much timber as the whole 4; what will be the measure of one side? *Ans.* 31·81+ in.

5. How many cubic feet in a log 50 feet long, and 16 inches in diameter? *Ans.* $69·8\frac{1}{3}$ cu. ft.

6. I have a joist 8 inches wide, and 3 inches thick. I want a stick of timber twice as wide, and containing $4\frac{1}{2}$ times as much timber; what must be its thickness? *Ans.* $6\frac{3}{4}$ in.

7. A man dug and stoned a well upon the following terms, viz.: for the first 15 feet he received $1·00 a foot, for the next 10 feet $1·50 a foot, for any greater depth $2·00 a foot, and for every foot of rock through

¶ **98.** Topic. Analysis. Applications of the principles contained in ¶¶ 52, 77, and 83. Rule, for a leaden ball; — for a cast-iron ball.

which he dug or blasted treble pay. He dug the well 35 feet deep, and from 22 to 29 feet through rock; how much did he receive for the job? *Ans.* \$71'00.

8. A laborer built a wall 5 rods long, 5 feet thick at the bottom, 2 feet thick at the top, and 5 feet high, in 2 days. He built 2 feet in hight of it the first day; on which day did he lay the most stone, and how much? *Ans.* 1st day, $8\frac{1}{4}$ cu. ft.

9. What is the area of a cylinder 14 feet long, and 8 feet 4 inches in circumference? *Ans.* 127'719+ sq. ft.

10. The diameter of the base of a cone is 8 feet, and the slant hight 32 feet; what is the area? *Ans.* 452'5894+ sq. ft.

11. A square pyramid 7 feet high, and 16 inches square at the base, sets on a pedestal 3 feet high, and 18 inches square; what was the cost of the whole, at \$1'$62\frac{1}{2}$ per cubic foot? *Ans.* \$17'709+.

12. A stick of timber 12 feet long, is 12 by 16 inches at one end, and runs to a point at the other; what is its solidity? *Ans.* $5\frac{1}{3}$ cu. ft.

13. The solidity of a square pyramid is 72 cubic feet, and one side of its base is 3 feet; what is its altitude? *Ans.* 24 ft.

14. The solidity of a cone is 8'83575 cubic feet, and the diameter of the base is 18 inches; what is its altitude? *Ans.* 15 ft.

15. A stick of timber is 31'5 feet long, the greater end 18 inches square, and the smaller end 8 inches square; what is its solidity? *Ans.* 38'$791\frac{2}{3}$ cu. ft.

16. The altitude of the frustrum of a cone is 9 feet, the diameter of the base 4 feet, and of the top 2 feet; what is its solidity?

17. The solidity of the frustrum of a cone is 65'9736 cubic feet, the diameter of the base 4 feet, and of the top 2 feet; what is the altitude?

18. I have ordered a marble frustrum of a cone, to be made of the following dimensions, viz.: diameter of base 4 feet, diameter of top 1'5 feet, and slant hight 8 feet; how much will it cost me, at \$2'$12\frac{1}{2}$ per cubic foot? *Ans.* \$106'577+.

19. A stick of hewn timber is $25\frac{1}{4}$ feet long, 20 inches square at one end, and 18 at the other; what are its contents in board measure? *Ans.* 760 sq. ft. 44 sq. in.

20. A cistern is 5 feet 10 inches in diameter at the top, 4 feet 8 inches at the bottom, and 6 feet deep; how many barrels of $31\frac{1}{2}$ standard gallons each will it contain? *Ans.* 30'9717 bar.

21. I wish to build a cistern that shall hold 918 gallons of water, and I would confine the diameter to 5 feet; what must be its depth? *Ans.* 6 ft. 3 in.

22. What is the side of the largest stick of square timber that can be hewn from a log 2 feet 4 inches in diameter? *Ans.* 19'79+ in.

23. What is the side of the largest stick of square timber that can be hewn from a log 35'5 inches in circumference? *Ans.* 7'99+ in.

24. A tree 18 feet long, and 40 inches in diameter, has 2 branches; one 10 feet long, and 10 inches in diameter; and the other 12 feet long, and 8 inches in diameter. The tree being of oak, an allowance of $\frac{1}{10}$ of the diameter is to be made for the thickness of the bark; what are the cubic contents of the tree and branches? *Ans.* 135'045603 cu. ft.

25. A log is 13 feet long, and 13 inches in diameter; how many feet of lumber of the standard thickness will it make, estimating by the second method? *Ans.* $195\frac{13}{20}$ sq. ft.

26. How many logs 14 inches in diameter, will make as much lumber of the standard thickness, as 100 logs 19 inches in diameter, estimating by the first method? *Ans.* 225.

27. What is the side of the largest cube that can be cut from a globe 2 feet in diameter? *Ans.* 13'854+ in.

28. Two pipes discharge water into a third, whose diameter is 13 inches, and the two together are sufficient to keep the third constantly full. One of the smaller pipes discharges twice as much water as the other in a given time; what is the diameter of each of the smaller pipes?
*Ans.* One is 7'5+ in.; the other 10'61+ in.

29. What are the superficies of a bomb 16 inches in diameter? What is the solidity? *Ans.* to last. 1 cu. ft. 416'6656 cu. in.

30. I want for my garden an iron roller, which shall be 20 inches in diameter, 4 feet 2 inches long, and the metal $1\frac{1}{2}$ inches thick. Allowing every cubic inch of the iron to weigh $4\frac{1}{2}$ ounces, what will the roller cost me, at \$'$06\frac{1}{4}$ per pound? *Ans.* \$76'$62\frac{1}{4}$+.

31. The globe or ball on St. Paul's church, in London, is 6 feet in diameter; what did the gilding cost, at 5 cents per square inch?
*Ans.* \$814'30.

32. A cast-iron ball weighs 72 pounds; what is its diameter?
*Ans.* 8 in.

33. The walls of a house are each 27'5 feet long, and 26 feet high, and the gables are 10 feet high. 10 feet high of the walls are $2\frac{1}{2}$ bricks thick, 10 feet more 2 bricks thick, 6 feet $1\frac{1}{2}$ bricks thick, and the gables 1 brick thick. What will the materials and bricklaying come to, at \$26'50 per square rod? (See ¶ 17, Note 2; and ¶ 18, Note 1.)
*Ans.* \$414'$19\frac{19}{96}$.

34. A stone house is 50 feet long, 34 feet wide, 20 feet high to the eaves, and the gables rise 9 feet above the eaves. 9 feet of the walls are 2 feet thick, the remainder to the eaves $1\frac{1}{2}$ feet thick, and the gables 1 foot thick. How many cubic feet in the walls? What did the building of them come to, at \$1'25 per perch, no allowance being made for windows, doors, &c.? *Ans.* to last. \$295'909+.

35. A man built a brick house 36 feet long, 28 feet wide, and 18 feet high to the eaves; and the walls were 1 foot thick. How many bricks did it take for the walls, allowing each brick to be 8 inches long, 4 inches wide, and 2 inches thick; and making no deductions for windows, doors, or mortar? *Ans.* 60264.

36. How many bricks would it take to build the above house, allowing $\frac{1}{4}$ of an inch upon three sides of each brick, for the thickness of the mortar? *Ans.* 48889+ bricks.

37. A laborer dug and stoned a cellar 21 feet long, 16 feet wide, and 7 feet deep, upon the following terms, viz.: he received \$'$18\frac{3}{4}$ per cubic yard for digging, \$'50 per perch for laying what stone he found in digging, and \$'$62\frac{1}{2}$ per perch for the remainder. In digging the cellar he found $11\frac{85}{99}$ perch of stone. The walls of the cellar being 2 feet thick, how much did he receive for his labor? (See ¶ 18.) *Ans.* \$41'012+.

38. The head diameter of a cask is 16½ inches, the bung diameter 21'6 inches, and the length 26 inches; what is its capacity in standard gallons? *Ans.* 35'007284 gal.

39. I have a granary 17 feet 6 inches long, 8 feet 9 inches wide, and 5 feet 3 inches deep. I wish to construct another whose dimensions shall be in a similar proportion, and which shall hold 8 times as much; what will be the dimensions of the new one, and what its capacity in bushels?

*Ans.* Dimensions, 35 ft. long, 17 ft. 6 in. wide, 10 ft. 6 in. deep; capacity, 5167'96⅞ bush.

40. A church spire is to be built in the form of a hexagonal pyramid, one side of the base being 10 feet, and the altitude 80 feet. Within the spire is to be a hollow cone 15 feet in diameter at the base, and so running up as to leave the walls of the spire as thick at the top of the cone as at the bottom. How many cubic feet will the spire contain?

*Ans.* 3393'902⅔ cu. ft.

# MECHANICAL POWERS.

¶ **100.** *The Mechanical Powers* are instruments or simple machines, employed to facilitate the moving of weights or the overcoming of resistance. They are six in number; viz., the *Lever*, the *Wheel and Axle*, the *Pulley*, the *Inclined Plane*, the *Wedge*, and the *Screw*.

In mechanical powers and in machinery, the thing to be moved, or the resistance to be overcome, is called the *Weight;* and the force which is applied to effect the object, is called the *Power*.

---

## The Lever.

¶ **101.** *The Lever* is an inflexible bar or rod supported at, and movable about a point.

*The Fulcrum* or *Prop* is the point upon which a lever is sustained, and about which it moves. The distances from the fulcrum to the points of the lever at which the weight and power act, are called the *Arms* of the lever.

Levers are commonly divided into three kinds, according to the relative positions of the power, the weight, and the fulcrum.

In a lever of the first kind the fulcrum F is between the power P and the weight W. AB is the long arm, and BC the short arm.

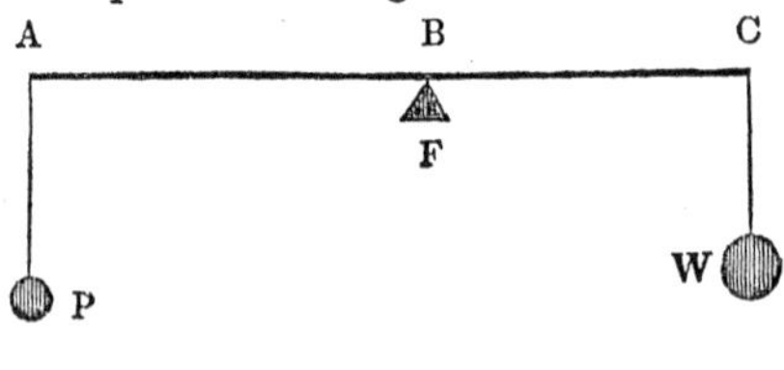


---

¶ **100.** Topic. Mechanical powers. Their number and names. Weight. Power.

¶ **101.** Topic. Lever. Fulcrum or prop. Arms of the lever. Kinds

In a lever of the second kind the weight W is between the power P and the fulcrum F. AC is the long arm, and BC the short arm.

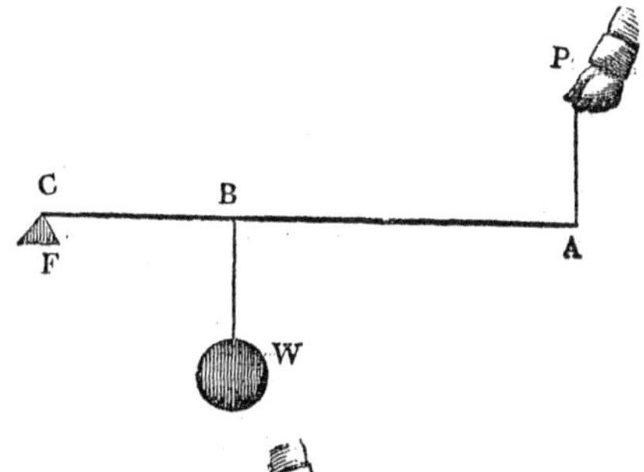


In a lever of the third kind the power P is between the weight W and the fulcrum F. BC is the long arm, and AC the short arm.

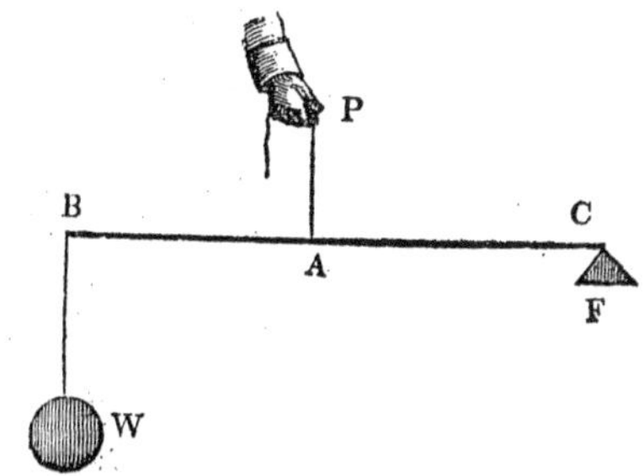


In making estimates of the force of the lever, four things are always considered; viz., the long arm, the short arm, the power, and the weight; any three of which being given, the other may be found.

**¶ 102.** It is a fixed principle in mechanics, that *a lever will remain at rest, when the ratio of the weight to the power is equal to the ratio of the long arm to the short arm.* In other words, *a lever will remain at rest, when the weight × by its distance from the fulcrum = the power × by its distance from the fulcrum.*

By letting $W$ represent the weight, $P$ the power, $LA$ the long arm, and $SA$ the short arm, the several processes may be conveniently presented in formulas, thus:

I. *The power, long arm, and short arm being given, to find the weight.*

$$\frac{P \times LA}{SA} = W.$$

---

of lever. 1st kind. 2d kind. 3d kind. Number of things always considered in making estimates upon the lever.

II. *The power, long arm, and weight being given, to find the short arm.*

$$\frac{P \times LA}{W} = SA.$$

III. *The weight, long arm, and short arm being given, to find the power.*

$$\frac{W \times SA}{LA} = P.$$

IV. *The weight, short arm, and power being given, to find the long arm.*

$$\frac{W \times SA}{P} = LA.$$

NOTE. The above formulas are equally applicable to either of the three kinds of levers.

### EXAMPLES FOR PRACTICE.

1. The long arm of a lever is 70 inches, the short arm 2 inches, and the weight 900 pounds; what power will be required to balance the weight?

2. A man has a lever 8 ft. 4 in. long, resting upon a prop 6 inches from one end. If he press upon the end of the longer arm with a force of 150 pounds, what weight at the end of the shorter arm will be required to balance him? *Ans.* 14100 lbs.

3. A lever 96 inches long has the fulcrum at one end, and a power of 50 pounds lifting at the other; what weight hung at 16 inches from the fulcrum, will be sufficient to keep the lever in a state of rest? *Ans.* 300 lbs.

4. A lever 9 feet long is fastened at one end, and has a weight of 187·5 pounds at the other; how far from the fulcrum must a power of 281·25 pounds be applied to sustain the weight? *Ans.* $36\frac{4}{9}$ in.

5. What power 70 inches from the fulcrum, will balance 900 pounds 2 inches from the fulcrum? *Ans.* $25\frac{5}{7}$ lbs.

6. The long arm of a lever is 48 inches, the power 5 pounds, and the weight 136 pounds; what is the length of the short arm?

**¶ 103.** *The ratio of the weight to the power is equal to the ratio of the velocity of the power to the velocity of the weight.* That is, *the power* × *by the distance through which it passes in a vertical direction* = *the weight* × *by the distance through which it passes in a vertical direction.*

---

**¶ 102.** The fixed principle of the lever. Abbreviations. 1st formula; 2d; 3d; 4th. Note.

**¶ 103.** Principle.

7. The weight is 28 pounds, and the power 8 pounds; what vertical distance does the weight pass through while the power ascends 2 feet?

*Ans.* $6\frac{6}{7}$ in.

8. The weight ascends 5 inches while the power descends $17\frac{1}{2}$ inches, and the weight is 30 pounds; what is the power? *Ans.* $9\frac{1}{7}$ lbs.

9. A lever is 72 inches long, and the fulcrum is so placed, that one end of the lever ascends a vertical distance of 18 inches, while the other descends a vertical distance of 30 inches. What is the length of each arm of the lever? What is the ratio of the weight to the power? What weight suspended from the end of the short arm will balance a weight of 80 pounds suspended from the end of the long arm?

*Ans.* to last. $133\frac{1}{3}$ lbs.

---

## The Wheel and Axle.

**¶ 104.** *The Wheel and Axle* consists of a wheel concentric with a cylindrical axis, with which it revolves; the power being applied to the circumference of the wheel, and the weight to that of the axis.

The wheel and axle is a perpetual lever, so contrived as to have a continued motion about a fulcrum. The radius of the wheel may be regarded as the long arm of the lever, and the radius of the axle the short arm. Hence,

**¶ 105.** *A wheel and axle will remain at rest, when the ratio of the weight to the power is equal to the ratio of the radius of the wheel to the radius of the axle.* That is, *when the weight* $\times$ *by the radius of the axle* $=$ *the power* $\times$ *by the radius of the wheel.* ¶ 102. And,

---

**¶ 104.** Topic. The wheel and axle. Its relation to the lever.

**¶ 105.** First principle; Second. Methods of applying power. Thickness of the rope. 1st formula; 2d; 3d; 4th.

*The ratio of the weight to the power is equal to the ratio of the circumference of the wheel to the circumference of the axle.* That is, *the power* $\times$ *by the circumference of the wheel* = *the weight* $\times$ *by the circumference of the axle.* ¶ 103.

The power is applied to the wheel and axle in various ways: sometimes by a rope; sometimes by pins which are grasped by the hand, as shown in the preceding diagram; and sometimes by a winch or crank, as seen in the common windlass. But in all cases the distance from the point at which the power is applied to the center of the wheel, is to be regarded as the radius of the wheel. When the thickness of the rope is considered, the force must be conceived as acting at the center of the rope, and therefore the thickness of the rope must be added to the diameter of the axle; and the thickness of that which supports the power, (if it be applied by a rope passing round the wheel,) must be added to the diameter of the wheel.

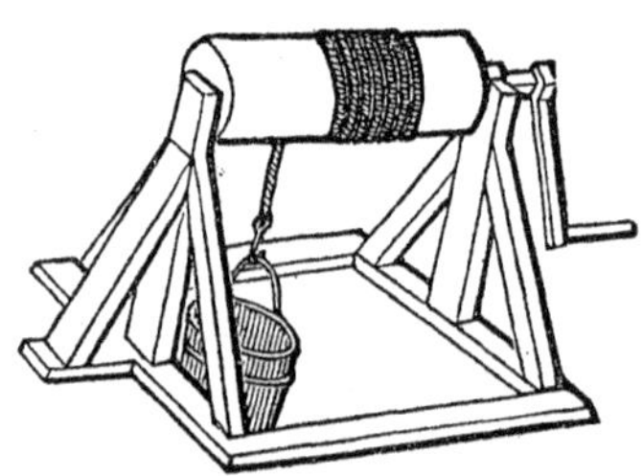

The formulas given for the lever, ¶ 102, may be adapted to the wheel and axle.

## EXAMPLES FOR PRACTICE.

1. The diameter of the wheel is 60 inches, and that of the axle is $7\frac{1}{2}$ inches; what weight upon the axle will be balanced by a power of 24 pounds at the circumference of the wheel? *Ans.* 192 lbs.

2. The diameter of a wheel is 35 inches, and of its axle 9 inches; what power applied at the circumference of the wheel, will balance a weight of 2240 pounds suspended by a rope 1 inch in diameter passing around the axle? *Ans.* $589\frac{9}{19}$ lbs.

3. The diameter of a wheel is 15 feet, the weight is 1000 pounds, and the power 3 pounds; what must be the diameter of the axle, that the weight may balance the power?

4. The diameter of an axle is 8 inches, the weight is 250 pounds, and the power 20 pounds; what must be the diameter of the wheel, that the weight may balance the power?

5. The diameter of a wheel is 8 feet, and of its axle 20 inches; what is the ratio of the weight to the power? What power will balance a weight of 872 pounds, the power and weight being each sustained by a rope $1\frac{1}{2}$ inches in diameter? *Ans.* to last. $192\frac{56}{195}$ lbs.

6. The weight is 100 pounds, the diameter of the axle 15 inches, and the weight ascends 1 foot while the power descends 4·8 feet; what is the diameter of the wheel? What is the power? *Ans.* to last. $20\frac{5}{6}$ lbs.

## The Pulley.

**¶ 106.** *The Pulley* consists of a wheel, movable about an axis, and having a groove cut in its circumference, over which a cord passes. The wheel is generally called a *sheave,* and is fixed in a box called a *block.*

NOTE. A pulley is frequently called a *tackle.*

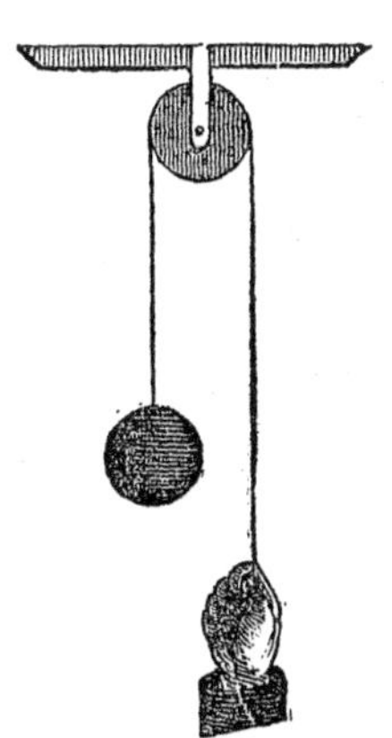

A single pulley affords no mechanical advantage, but serves merely to change the direction of the motion. No mechanical advantage is gained from any number of fixed pulleys.

Two or more pulleys, one at least of which is movable, may be combined in various ways, by which a mechanical advantage is gained, greater or less, according to the number of movable pulleys, and the mode of their combination. Thus, the weight W is supported by the two parts of the cord which passes under the pulley E, $\frac{1}{2}$ being sustained by each part. Consequently, the power P must be $\frac{1}{2}$ as great as the weight W, in order to balance it.

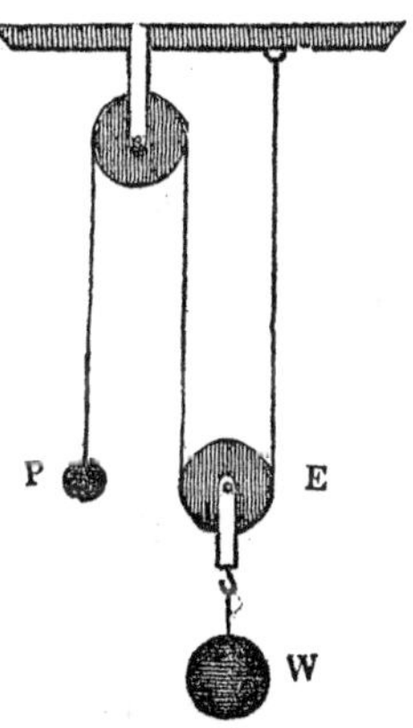


---

¶ **106.** Topic. Pulley. Sheave. Block. Note. Use of single pulley. Combination of pulleys. Blocks or systems of pulleys. Smeaton's pulley.

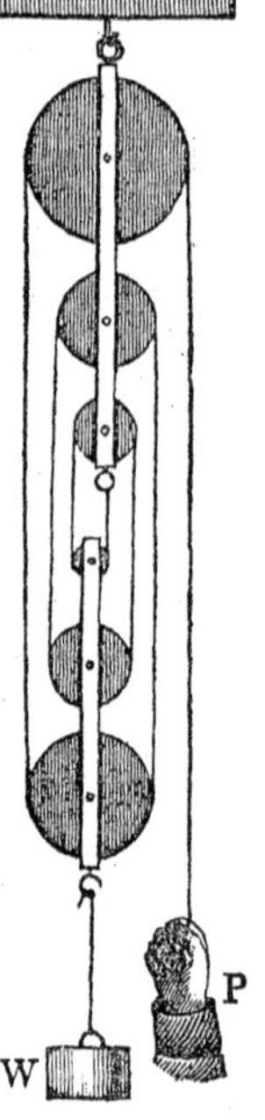


The accompanying diagram represents *Blocks of Pulleys*, also called a *System of Pulleys*. The weight W is supported by the 6 parts of the rope which passes under 3 movable, and over 3 fixed pulleys. Since 6 parts of the rope support the whole weight, 1 part must support $\frac{1}{6}$ of it. Therefore, the power P is $\frac{1}{6}$ as great as the weight W.

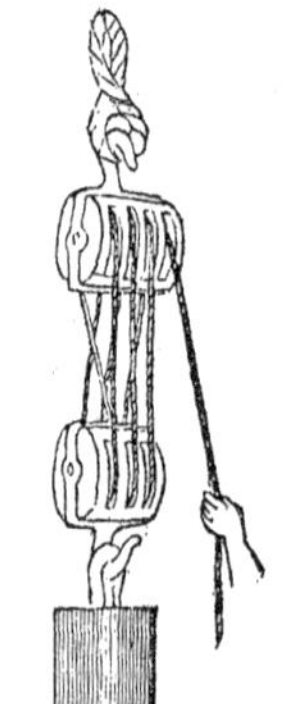

The combination of pulleys shown in this diagram, though differing somewhat from the last in form, is the same in effect, consisting of 3 movable, and 3 fixed pulleys. But this combination is liable to some objections, the most important of which is, that unless the weight is guided by the hand, the ropes will twist together. This objection led to the invention of

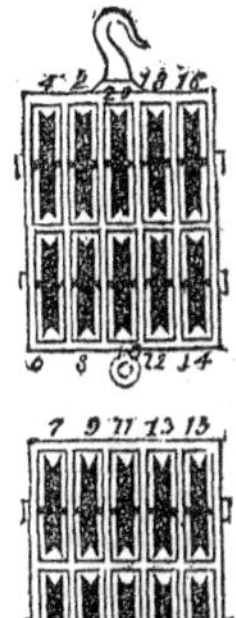

*Smeaton's Pulley*, or *Smeaton's Tack*, as it is usually called. This contains 2 rows of wheels in each block, as shown in the diagram. A single cord is made to pass over them, in such a manner, that the power and the weight both act in the same line with the centers of the 2 blocks, thus preventing the tendency to twist. The pupil will readily trace the pulleys, in the order that the rope passes over them.

**¶ 107.** *A pulley will remain at rest, when the power is to the weight as* 1 *is to the number of ropes; or, as* 1 *is to twice the number of movable pulleys.* That is, *the power* × *by the number of ropes, or by twice the number of movable pulleys* = *the weight.* And,

*The ratio of the weight to the power is equal to the ratio of the distance through which the power moves to the distance through which the weight moves.* That is, *the weight* × *by the distance through which it moves* = *the power* × *by the distance through which it moves.*

The formulas given for the lever, ¶ 102, may be adapted to the pulley.

### EXAMPLES FOR PRACTICE.

1. The number of movable pulleys is 4, and the power is 32 pounds; what weight will be required to balance the power? *Ans.* 256 lbs.

2. The weight is 721 pounds, and the number of ropes is 8; what power will be required to balance the weight? *Ans.* $90\frac{1}{8}$ lbs.

3. The weight is 250 pounds, and the power $12\frac{1}{2}$ pounds; what is the number of movable pulleys? What the number of ropes?
*Ans.* to last. 20 ropes.

4. The weight is 900 pounds, and ascends 3 inches while the power descends 30 inches; what is the number of movable pulleys? What is the power? *Ans.* to last. 90 lbs.

---

¶ **107.** First principle; Second. 1st formula; 2d; 3d; 4th.

## The Inclined Plane.

**¶ 108.** *The Inclined Plane* is a hard plane surface, forming an acute angle with a horizontal plane.

When a body is moved upwards vertically, its entire weight must be overcome by the power; but when it is moved up an inclined plane, a twofold effect is produced. A part of the weight is sustained by the plane, and the remainder presses against any surface which would resist its motion down the plane. Thus, in the accompanying diagram, a portion of the weight is sustained by the plane, and the remainder by the power.

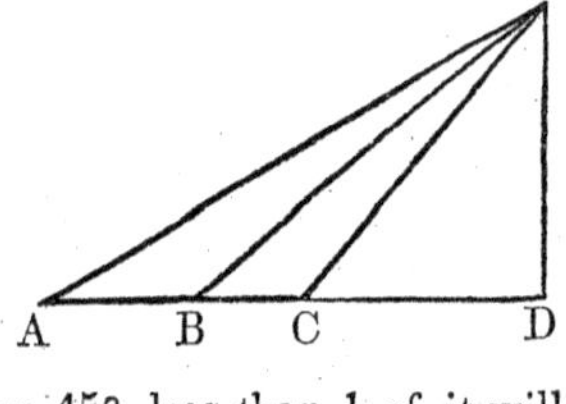


If a weight be moved along the horizontal plane AD, it will be supported by the plane; if it be moved up the vertical plane DE, it will be supported entirely by the power; if it be moved up the inclined plane BE, at an angle of 45°, it will be supported $\frac{1}{2}$ by the plane, and $\frac{1}{2}$ by the power; if it be moved up the plane CD, at an angle of more than 45°, less than $\frac{1}{2}$ of it will be supported by the plane, and the remainder by the power; and if it be moved up the plane AE, at an angle of less than 45°, more than $\frac{1}{2}$ of it will be supported by the plane, and the remainder by the power. And universally,

*The power necessary to support any given weight upon an inclined plane, depends upon the length of the plane, and the abruptness of its ascent.*

**¶ 109.** Hence, *The power is to the weight as the hight of the plane is to its length.* That is, *the weight* × *by the hight of the plane* = *the power* × *by its length.*

The formulas given for the lever, ¶ 102, may be adapted to the inclined plane.

---

¶ **108.** Topic. Inclined plane. Analysis. Different angles of inclined planes. Principle.

¶ **109.** Principle. 1st formula; 2d; 3d; 4th.

**EXAMPLES FOR PRACTICE.**

1. The length of an inclined plane is 12 feet, and its hight is 2 feet; what power will be necessary to support a weight of 200 pounds upon the plane? *Ans.* $33\frac{1}{3}$ lbs.

2. A hill rises 440 feet in half a mile; what weight will a man, pulling with a force of 150 pounds, be able to keep from rolling down the hill? *Ans.* 900 lbs.

3. A boy, by bracing with a force of 70 pounds, is able to hold a barrel of oil, weighing 400 pounds, upon an inclined plane 15 feet long; what is the hight of the plane? *Ans.* $2\frac{5}{8}$ ft.

4. A power of 90 pounds will hold a weight of 2700 pounds upon an inclined plane 15 feet high; what is the length of the plane? *Ans.* 450 ft.

---

## The Wedge.

**¶ 110.** *The Wedge* consists of an inclined plane, or of two inclined planes joined together, the entire length of their bases.

It is sometimes used for raising bodies, by being made to pass under them; but more frequently for dividing or splitting them. In the former case, if we suppose the wedge to be pushed under the load by pressure, its action is precisely the same as that of the inclined plane; for the effect is the same, whether the wedge be pushed under the load, or the load be drawn up the plane.

But the wedge is more commonly driven forward, by blows from a mallet or hammer, while the resistances which it has to overcome act with constant force against it. Hence, its power cannot be estimated with any degree of accuracy. It may, however, be stated, that *the mechanical advantage of the wedge is increased by diminishing the angle of its cutting edge; but the strength of the tool is thereby diminished.*

All cutting and piercing instruments, such as axes, knives, scissors, chisels, &c., nails, pins, needles, awls, &c., are modifications of the wedge. The angle of the cutting edge of the wedge is made more or less acute, according to the purpose to which it is to be applied. In tools for cutting wood, the angle is generally about 30°; for iron, it is from 50° to 60°; and for brass, from 80° to 90°. In general, the softer the substance to be divided is, the more acute may be the angle of the wedge; and tools which act by pressure may have their cutting edges more acute than those which are driven by a blow.

---

**¶ 110.** Topic. Wedge. Its uses. Difficulty in estimating the power of the wedge. Principle. Examples of the wedge. Angle of the cutting edge.

## The Screw.

¶ **111.** *The Screw* consists of a spiral ridge or groove, winding round a cylinder, so as to cut every line on the surface parallel to the axis at the same angle.

If the inclined plane AC be wound round a cylinder whose circumference equals the base AB, the plane AC will form the thread of a screw; and, if the plane be continued, the perpendicular BC will be the distance between any two contiguous threads.

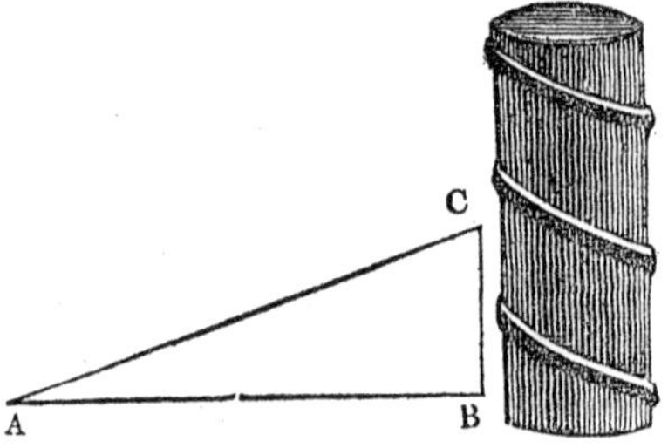


Hence, The screw may be supposed to consist of an inclined plane, whose base is the circumference of the screw, and whose altitude is the perpendicular distance between any two contiguous threads.

In the application of the screw, the weight is not placed upon the threads, but the power is transmitted by causing the screw to move in a hollow cylinder, whose interior surface contains a spiral cavity corresponding to the thread of the screw, and in which the thread will move by turning round the screw continually in one direction. This hollow cylinder is usually called the *Nut*, or *Concave Screw.*

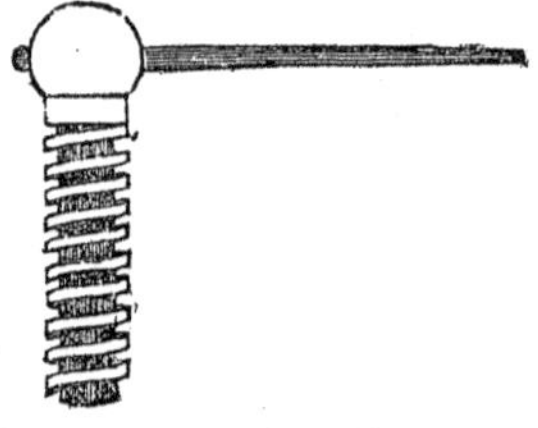

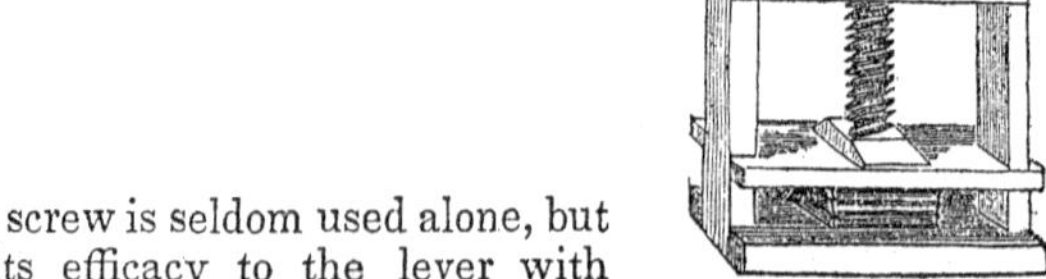

The screw is seldom used alone, but owes its efficacy to the lever with

¶ **111.** Topic. Screw. Its relation to the inclined plane. Application of the screw. Nut. Manner of applying the power. Movable screw. Movable nut.

which it is connected. The lever is sometimes connected with the screw, and sometimes with the nut. When connected with the screw, the nut is immovable; and when with the nut, the screw is immovable.

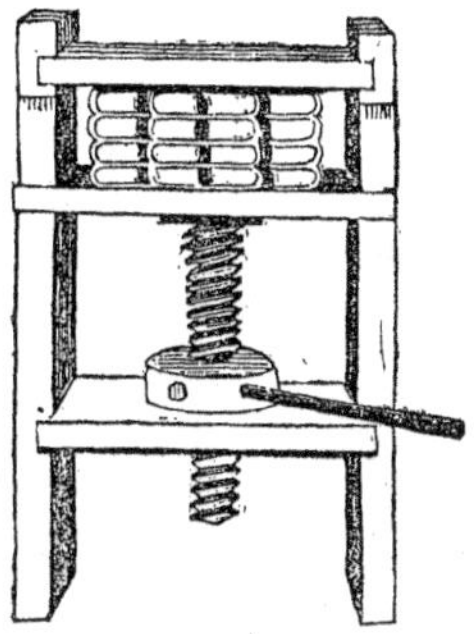

¶ **112.** *A screw will remain at rest, when the ratio of the power to the weight is equal to the ratio of the distance between the adjacent threads of the screw to the circumference described by the point to which the power is applied.* That is, *the weight* $\times$ *by the distance between the adjacent threads* = *the power* $\times$ *by the circumference of a circle whose radius is equal to the length of the lever.*

Hence, the mechanical advantage afforded by the screw is dependent upon the fineness of the threads, the smallness of the cylinder or body of the screw, and the length of the lever by which the power is applied.

The formulas given for the lever, ¶ 102, may be adapted to the screw.

### EXAMPLES FOR PRACTICE.

1. The threads of a screw are 1 inch asunder, the length of the lever by which it is turned is 3½ feet, and the power is 30 pounds; what weight will be necessary to balance the power? *Ans.* 660 lbs.

2. The distance between the threads of a screw is 1¼ inches, and the length of the lever is 21 inches; what power is necessary to balance a weight of 9900 pounds? *Ans.* 112·5 lbs., nearly.

3. The weight is 25000 pounds, the power 150 pounds, and the circumference of the circle described by the power 10 feet; what must be the distance between the threads of the screw, that the weight may balance the power? *Ans.* ·72 of an inch.

4. The threads of a screw are ¼ of an inch apart, the weight is 362057¼ pounds, and the power 120 pounds; what must be the length of the lever by which the power is applied, that the power will balance the weight? *Ans.* 14·41+ in.

---

¶ **112.** Principle. Mechanical advantage of the screw. 1st formula; 2d; 3d; 4th.

## Friction.

**¶ 113.** *Friction* is the resistance produced by the rubbing of the surfaces of two solid bodies against each other.

If the surfaces of bodies were perfectly smooth and polished, they would slide upon one another without any resistance from their contact. But this state of smoothness and polish never exists, and can never be attained. The surfaces of all bodies, even when they have received the highest polish that we are capable of giving them, retain a greater or less degree of roughness, which prevents them from sliding upon one another without resistance or friction.

Friction is of three kinds; viz.:

1st. That which occurs when one body slides upon the surface of another.

2d. That of rolling bodies; and,

3d. That of the axles of wheels.

**¶ 114.** From the numerous and varied experiments upon this subject, have been deduced the following conclusions.

I. *The friction of sliding bodies.*

1. Between similar substances, under similar circumstances, friction is a constant retarding force.

2. Friction is greatest between bodies whose surfaces are rough, and is lessened by polishing them.

3. It is greater between surfaces of the same material, than between those composed of different materials.

4. If the rubbing surfaces remain the same, the friction increases directly as the pressure.

5. If the pressure remain the same, the friction has no relation to the extent of the surface.

6. The application of oil, grease, or any unguent, in general diminishes the friction, though in different degrees, dependent upon various circumstances.

**¶ 115.** II. *The friction of rolling bodies.*

1. Like the friction of sliding bodies, it is a constant retarding force.

---

¶ **113.** Topic. Friction. Causes of friction. 1st kind of friction; 2d kind; 3d kind.

¶ **114.** First law of the friction of sliding bodies; 2d law; 3d; 4th; 5th; 6th.

¶ **115.** First law of the friction of rolling bodies; 2d law; 3d; 4th; 5th; 6th.

2. It is affected by the nature of the surface, so far as polish is concerned; but is not lessened by the application of unctuous substances.

3. It is less between bodies of different materials, than between those of the same substance.

4. It is directly proportional to the pressure.

5. It has no relation to the extent of the surface.

6. It is much less in rolling than in sliding bodies.

**¶ 116.** III. *The friction of the axles of wheels.*

1. It is less than that of sliding, but greater than that of rolling bodies.

2. It follows in all respects the laws of the friction of sliding bodies.

3. A great advantage may be obtained from greasing the surfaces. By the application of fresh tallow, the friction is reduced about one half.

**¶ 117.** No definite rules for the allowances which must be made for friction have yet been established. The following allowances are as nearly correct as any that have been presented.

| | | | | |
|---|---|---|---|---|
| The friction of sliding bodies is from ·12 to ·35 of the weight or pressure, or | from 12 | to | 35 | per cent. |
| The friction of rolling bodies, | " 5 | " | 12 | " |
| " " " the axles of wheels, | " 12 | " | 20 | " |

The friction of the axles of wheels may be stated more definitely, as follows:

| | | | | |
|---|---|---|---|---|
| An iron axle turning in a box of oak, | from 12 | to | 18 | per cent. |
| A wooden axle turning in a box of wood, | " 8 | " | 15 | " |
| A metallic axle turning in a box of another metal, and well coated with grease, | " $2\frac{1}{2}$ | " | 6 | " |
| A wooden axle turning in a box of wood, and coated in a similar manner, | " 3 | " | 8 | " |
| An iron axle turning in a box of wood, and coated in a similar manner, | " 5 | " | 10 | " |

In making allowances for friction, the per cent. of allowance is to be estimated upon, and added to the weight or resistance, or estimated upon, and subtracted from the power, before any other computations are made.

---

¶ **116.** First law of the friction of the axles of wheels; 2d law; 3d.

¶ **117.** Rates per cent. of the friction of sliding bodies. Of rolling bodies. Of the axles of wheels. More definite statement of the friction of the axles of wheels. How allowances for friction are to be made.

## General Remarks upon the Mechanical Powers.

**¶ 118.** The mechanical powers afford no positive gain in any one respect, that is not counterbalanced by a loss in another. Thus, one man, with the aid of one of the simple mechanical powers, is able to remove a weight, or to overcome a resistance, that would require the united strength of 20 men unaided by any machine. Here is a positive gain in one respect. But, to effect this end, the one man must cause the power which he applies to move 20 times as far as he wishes to move the weight; and the time required to perform the labor will be 20 times as great as would be required, had 20 men been employed. Here, then, is a loss in distance passed over by the one man, and in the time consumed, fully equal to the advantage afforded by the machine employed.

**¶ 119.** The friction in the several mechanical powers is various, and is dependent upon the form of the machine and the materials of which it is composed. Thus, the friction of a lever, when turning upon a sharp-edged fulcrum of hardened steel, is so very small that, in ordinary cases, it is not taken into account; while, if the fulcrum be a stone or a block of wood, the friction may be as much as 3 per cent. The friction of the wheel and axle is modified by various circumstances, which have been named in ¶ 117. But, when cordage is employed, it increases the resistance from 7 to 10 per cent. In the pulley the friction of the cordage, together with that of the sheaves and blocks, increases the resistance from 20 to 75 per cent. The friction of bodies rolling on inclined planes, is usually so small, that in estimates it is not considered. That of sliding bodies has already been given, ¶ 117. The friction of the screw is very great; it must exceed the resistance, or the screw will not retain its position. Screws with sharp or wedge-shaped threads are attended with more friction than those whose threads are square.

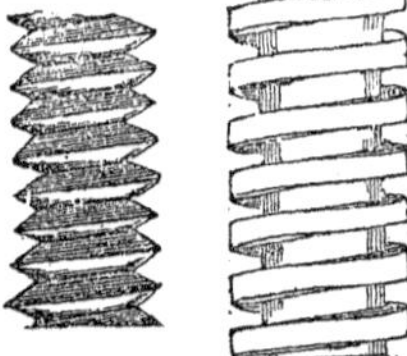

---

¶ **118.** Topic. Advantage of the mechanical powers. Illustration.

¶ **119.** Causes of the difference in the friction of the different mechanical powers. Friction of the lever. Of the wheel and axle. Of the pulley. Of the inclined plane. Of the screw.

# MACHINERY.

¶ **120.** Any machine, however simple or complex in its construction and operation, must contain one or more of the simple mechanical powers; nor can it involve any other principles than those of the mechanical powers.

*A Simple Machine* is one which involves but one of the simple mechanical powers.

*A Compound Machine* is formed by combining two or more simple machines.

## ¶ 121. Methods of Transmitting Motion.

Motion may be transmitted from the moving power to the other parts of a machine in various ways, dependent upon circumstances. When two parts of a machine, acting at some distance from each other, are to be moved together, in the same direction, the motion may be transmitted by a band, passing over wheels attached to the two parts of the machine. And when the two parts to be connected are to move in contrary directions, the band may be crossed. If a rope band be used, its friction, and consequently its efficacy, may be increased, by grooving the edge of the wheels. And when a strap band is used, its friction may be increased, by increasing the width of the band.

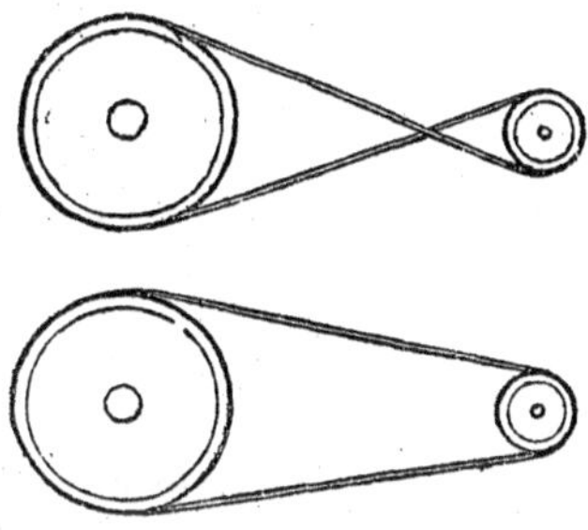

¶ **122.** But the more common method of transmitting motion is, by wheels having *Teeth* or *Cogs* cut in their circumference. The connection of toothed wheels with each other,

---

¶ **120.** Topic. Principles involved in every machine. Simple machine. Compound machine.

¶ **121.** Topic. Bands. Cross bands. Friction of bands.

¶ **122.** Gearing. Pinion. Its leaves. Kinds of toothed wheels. Spur wheel. Spur gearing. Crown wheel. Its effect when working with a spur wheel. Bevel wheel. Bevel gearing. Use of bevel wheels. Note. Universal joint.

for the purpose of transmitting motion in machinery, is called *Gearing*. It is usual to call a small wheel acted upon by a large one, a *Pinion*, and its teeth the *Leaves* of the pinion.

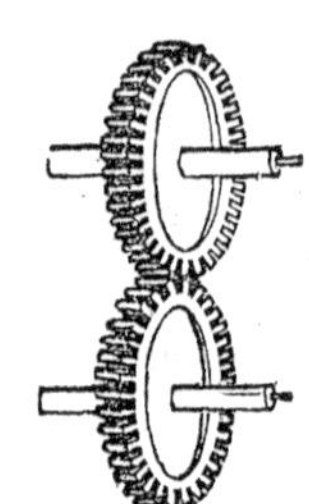

Toothed wheels are of three kinds; viz.:

1st. When the teeth are raised upon the edge of the wheel, or are perpendicular to the axis, the wheel is a *Spur Wheel;* and two or more spur wheels working together are called *Spur Gearing*.

2d. When the teeth are raised parallel to the axis, or perpendicular to the plane of the wheel, it is called a *Crown Wheel*. A crown and a spur wheel working together, serve to transmit the motion of one to the other at a right angle.

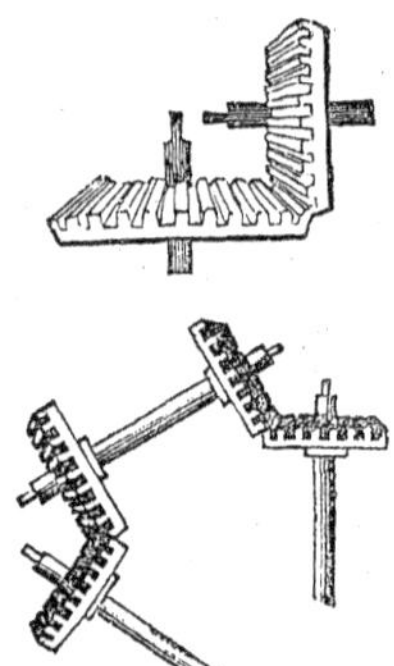

3d. When the teeth are raised on a surface inclined to the plane of the wheel, it is called a *Beveled Wheel;* and two or more beveled wheels working together are called *Beveled Gearing*. Beveled wheels are employed to transmit motion from one axis to another inclined to it, at any proposed angle.

NOTE. Beveled wheels are also called *conical wheels*, because their teeth may be regarded as cut in the frustrum of a cone.

The direction of motion may be changed from a right line to any angle less than 40°, by the *Universal Joint.* This is effected by connecting the ends of two axes with the joint, as shown in the diagram.

**¶ 123.** *Gudgeons*, in machinery, are pins inserted in the extremities of a shaft, or the axle of a wheel, on which it turns, and which support the weight. In order to diminish friction, gudgeons are made as small as possible in diameter; leaving, however, sufficient strength to support the weight.

*The Box* of a gudgeon is the hollow cylinder of wood or metal, in which the gudgeon runs.

---

## Teeth of Wheels.

**¶ 124.** Where the teeth of wheels work into each other, as represented in the diagram, every part of the side of each tooth of one wheel comes successively in contact with a tooth of the other, as the wheel turns round, and consequently the force is exerted at the points which are in contact. But it is of the utmost importance that the parts act upon each other with a uniform force, and with the least possible amount of friction. This end can be attained

---

¶ **123.** Gudgeons. Their size. Box of a gudgeon.

¶ **124.** Topic. Object in making the teeth of wheels curving. Object in making the number of teeth of two wheels, or of a wheel and pinion working together, prime to each other. Illustration. Hunting-cog.

in no other way than by making the teeth of the wheels curving. The curve of the teeth will be greater or less, according to the size of the wheel and the dimensions of the teeth.

But the surfaces of teeth will always contain some inequalities, and consequently will cause some friction. To equalize the wear arising from inequalities on the surface of the teeth of wheels and pinions, each tooth of the pinion should work in succession in every tooth of the wheel, and not always in the same set of teeth. To effect this, the number of teeth in a wheel and in a pinion which work into each other, must be prime to each other. Thus, if the wheel contain 61 teeth, and the pinion 12, each tooth of the pinion will work in succession in every tooth of the wheel. In this case, no tooth of the pinion can act with the same tooth of the wheel a second time, until it has acted upon every other tooth of the wheel. The odd tooth which produces this effect, is called the *Hunting Cog*.

---

## Horse Power.

**¶ 125.** The force of a machine or an engine may be obtained and applied in a variety of ways, as by gravity, animal strength, wind, water, steam, &c.; but in estimating the power of any machine of great force, the power is referred to a fixed and established standard, called *Horse Power*.

*Horse Power* is the weight which a horse is capable of raising to a given hight in a given time. Custom has established as a standard, that a machine of one horse power is capable of raising a weight of 33000 pounds one foot in a minute.

NOTE. A machine of 1 horse power will raise a weight of 2000 lbs. 1 rod in a minute; 500 lbs. 4 rds. in a min.; and 125 lbs. 16 rds. in a min., or 3 miles an hour.

---

**¶ 125.** Topic. Ways in which the force of a machine may be obtained. Fixed and established standard to which the power of any machine of great force is referred. Horse power. A machine of one-horse power. Note.

## Levers and Weighing Machines.

**¶ 126.** Any number of weights may be attached to either arm of a lever. *The lever will remain at rest, when the sum of the products of the weights upon one arm × by their respective distances from the fulcrum = the sum of the products of the weights upon the other arm × by their respective distances from the fulcrum.* See ¶ 102.

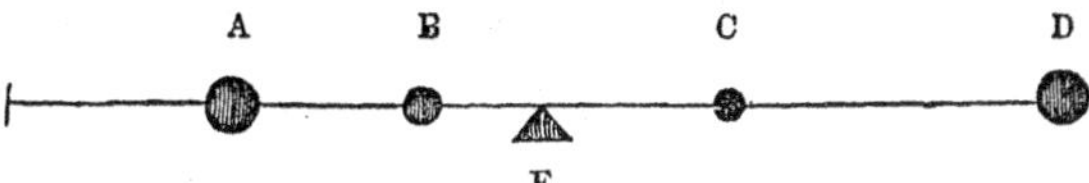


Ex. The distance from E to B is 4 inches, from B to A 5 inches, from E to C 5 inches, and from C to D 8 inches; A weighs 20 pounds, B 8 pounds, and C 3 pounds. What must be the weight of D, that the lever may remain at rest?

*Ans.* $15\frac{2}{13}$ lbs.

**¶ 127.** When several simple levers act upon each other, the combination is called a *Compound Lever*. The principles given for estimating the force of a simple lever, ¶ 102, are equally applicable to the compound lever. But, by a careful examination of the operations necessary to estimate the force of a compound lever, the pupil will find, that *a compound lever will remain at rest, when the product of all the arms on the side of the power × by the power = the product of all the arms on the side of the weight × by the weight.*

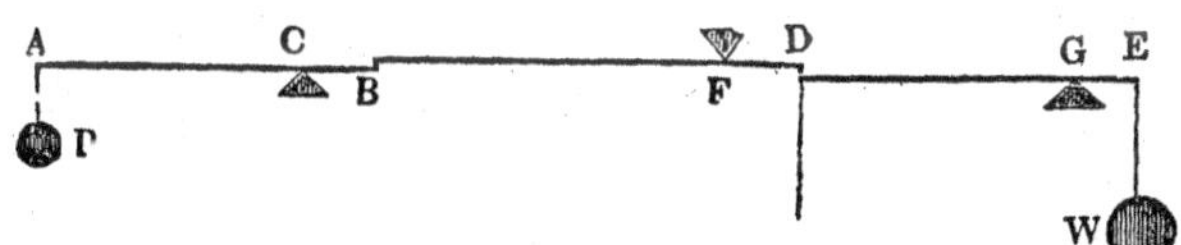


Ex. In the above compound lever AC is 9 in., BF 12 in., DG 10 in., BC 2 in., DF $2\frac{1}{2}$ in., and EG 2 in.; what weight suspended at E will balance a power of 15 pounds suspended at A?

*Ans.* 1620 lbs.

---

**¶ 126.** Topic. Principle.
**¶ 127.** Compound lever. Principle.

¶ **128.** *The Balance* consists of a beam or lever suspended exactly in the middle, with scales or basins hung at or suspended from the extremities, of precisely equal weight. The accuracy of the balance depends upon the length of its arms, and the shape and material of its fulcrum.

A fraudulent balance may be made, by making one arm of the scale beam shorter than the other. The fraud may readily be detected, by weighing an article in one scale, and then in the other. If it weigh the same in both, the balance is correct; otherwise, it is fraudulent.

*The actual weight of a body may be obtained by a false balance, as follows:*

1st. Weigh the body in the two scales successively.

2d. Multiply the two weights together, and extract the square root of their product.

NOTE. The square root of their product is a *Geometrical Mean* between the two weights. See Revised Arith., ¶ 185, note.

Ex. A body, when placed in one scale of a balance, weighs $8\frac{1}{2}$ pounds, but when placed in the other, it weighs 14 pounds; what is its true weight? *Ans.* 10·908+ lbs.

¶ **129.** *The Steelyard* is a balance, which consists of a lever having two unequal arms; the weights of bodies being determined by means of a single standard weight.

The body whose weight is to be determined, is suspended from the extremity of the short arm; and, in weighing, the constant weight or *Counterpoise* (commonly called *poise*,) is moved along the longer arm, until the lever is brought to rest in a horizontal position. Divisions marked on the longer arm, indicate the weight of any body suspended from the shorter arm, balancing the poise at any division.

Ex. I have a steelyard, a weight of 10 pounds, and a poise of $\frac{1}{2}$ pound. How will I proceed to lay off pound notches upon the long arm of the steelyard?

---

¶ **128.** Balance. Its accuracy. A fraudulent balance. Manner of detecting it. How to find the actual weight of a body, by a fraudulent balance. Note.

¶ **129.** Steelyard. Manner of using it.

**¶ 130.** When the power and the weight do not act on the lever in directions perpendicular to its length, or when the lever is bent or crooked, the perpendicular distances from the fulcrum to the lines of direction in which the power and weight act, are to be regarded as the arms of the lever. Thus, in the present position of the *Bent Lever Balance* represented in the diagram, BD and BK are to be considered as the arms of the bent lever CBK. The pupil will perceive that a small weight in the scale will elevate the weight C but a little distance upon the graduated scale FG. But weights may be added, till C shall be elevated to G. Every change of weight changes the relative distances of the power and weight from the fulcrum.

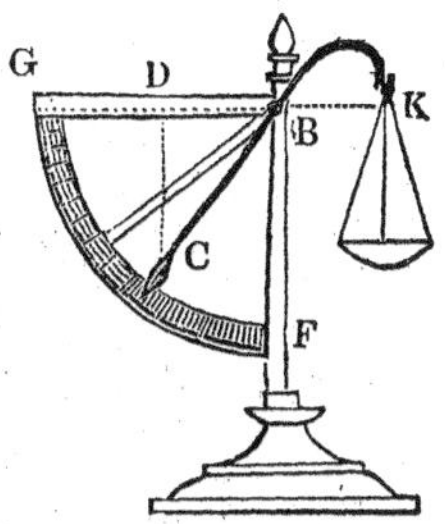


## Wheel Work.

**¶ 131.** *The Capstan* is a strong massive piece of timber, in the form of a cylinder or frustrum of a cone, around which a rope is coiled; and being turned by means of bars or levers, inserted into its head, or a drum attached to its head, it affords an advantageous mode of applying power to overcome resistance. The capstan is chiefly employed in ships for weighing anchors, hoisting sails, &c.; and on land, for moving buildings, &c. When used for the last-named purpose, it is commonly moved by horse power.

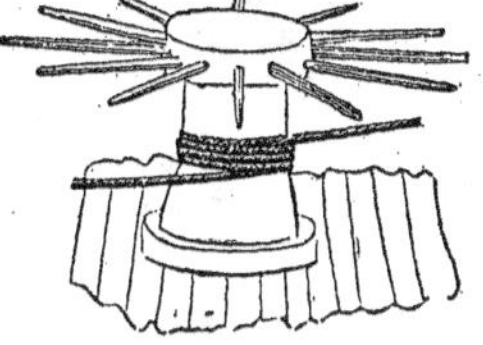

The power of the capstan may be greatly increased, by connecting with it an arrangement of wheel work.

### PRACTICAL EXAMPLES IN WHEEL WORK.

1. A capstan is 1 foot in diameter, the levers by which it is turned are each 6 feet long, and the rope to which the weight is attached is 3 inches in diameter; allowing 10 per cent. for the friction of the capstan, and $1\frac{1}{2}$ per cent. for the stiffness of the rope, what power must be applied by each of 5 men, at the end of the levers, to move a weight of 12000 pounds? *Ans.* 278·75 lbs.

---

¶ **130.** Principle of the bent lever. Illustration by the bent lever balance.

¶ **131.** Topic. Capstan. Where employed. Manner of increasing its power.

2. The lever of a capstan 2 feet in diameter is 12 feet long, and the rope by which the weight is moved is 2 inches in diameter; allowing 12 per cent. for the friction of the capstan, and 1 per cent. for the stiffness of the rope, what weight will be moved by a horse attached to the end of the lever, and pulling with a force of 900 pounds? *Ans.* $8673\frac{3}{13}$ lbs.

3. In the spur gearing represented in the diagram, the respective diameters of the wheels A, B, and C, are 14, 16, and 18 inches; and the diameters of their pinions a, b, and c, are 3, 4, and 5 inches; allowing 5 per cent. for the friction of the axles, and 3 per cent. for that of the teeth of the wheels and pinions, what power applied at P will be required to move a weight of 2000 pounds suspended at W?

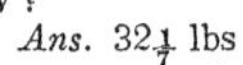

*Ans.* $32\frac{1}{7}$ lbs

4. The respective circumferences of the wheels A, B, C, D, and E, are 30, 22, 30, 35, and 44 inches; and of their pinions a, b, c, d, and e, 10, 10, 10, 11, and 12 inches; through what distance will the power P move, while the weight W moves 1 foot? *Ans.* 231 feet.

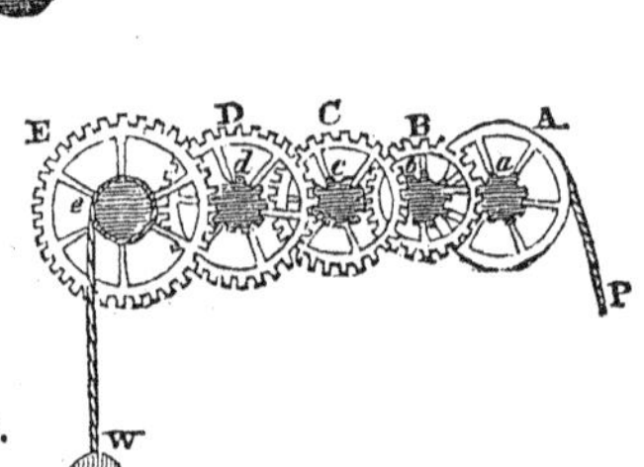


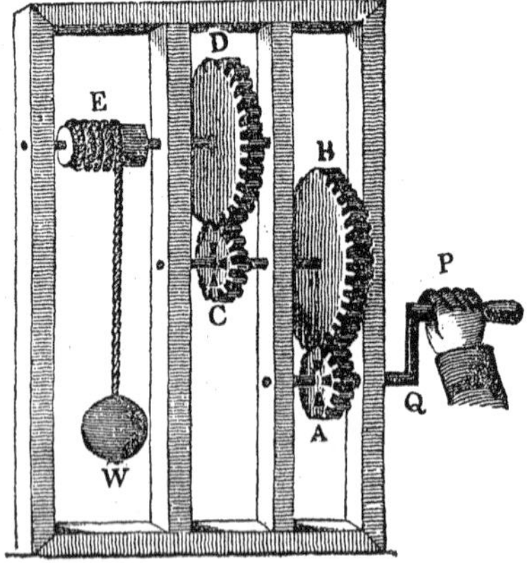


5. The wheels B, and D, are each 10 inches in diameter; the pinions A, and C, each 3 inches; the axle E 2 inches; the circumference of the circle described by the power P is 33 inches; and the rope which sustains the weight is 1 inch thick. Allowing the whole friction of the machine to be 9 per cent., what power applied at P, will be required to raise a weight of 6000 pounds suspended at W?

*Ans.* $168\frac{6}{35}$ lbs.

6. A man whose weight is 150 pounds, attempts to draw himself up, by a rope passing over a single fixed pulley. Allowing the friction of the axle of the pulley and of the rope and pulley to be 20 per cent., with what force must he pull upon the rope to effect his object ?

7. The weight A is 500 pounds, the friction of the pulley and the rope at B is 12 per cent., and of the pulley and rope at C 13 per cent.; what strength must the horse exert to raise the weight?
*Ans.* 625 lbs.

## White's Pulley.

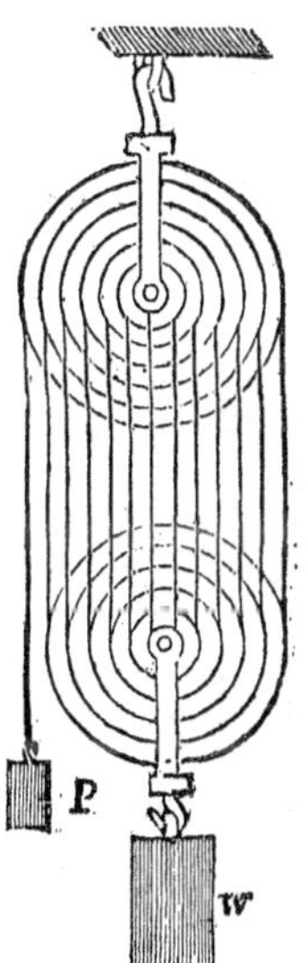


**¶ 132.** The great amount of friction offered by the forms of pulley that have been presented, renders their use in some measure objectionable. The friction of the sheaves and blocks, together with that of the cordage and the axles of the sheaves, are sometimes so great, as to render the pulley of no advantage. But these objections are removed in the pulley here presented, and known as *White's Pulley*. The wheels in each block turn on the same axis, and consequently revolve in the same time; and, instead of separate wheels, the upper and lower blocks are each cut in grooves in one block, thus reducing the friction of the sheaves and blocks, and of the axles, to that of one wheel in each block. The size of each wheel is so proportioned to the others, that any point in its circumference moves with the velocity of the rope on that wheel. To effect this, the diameters of the wheels in the upper block must be as the numbers 1, 3, 5, &c., and in the lower block as 2, 4, 6, &c.

**¶ 132.** Objections to the common forms of pulley. Explanation of

Ex. The weight W, in the diagram, is 1200 pounds, and the resistance offered by the friction of the pulley is 15 per cent.; what power applied at P, will be necessary to raise the weight? See ¶ 107. *Ans.* 115 lbs.

---

## The Crane.

**¶ 133.** *The Crane* is a machine for raising heavy weights, and depositing them at some distance from their original place. Its parts are a *jib* or transverse beam CD, inclined to a perpendicular in an angle of 40° or 50°. This is connected with the *vertical beam* AB, which is fastened to the floor, but is capable of turning on its axis. The upper end of the jib carries a *fixed pulley* at D, over which passes a *rope* or *chain*, with a *hook* at O to support the weight. The wheel-work is mounted in two cast-iron crosses attached to the beam AB, one of which is shown at EF–GH. I is the winch at which the power is applied. This carries a pinion which works in the wheel K; a pinion upon the axle of the wheel K, works in the wheel L; and upon the axle of the wheel L, is a cylinder or barrel, on which the rope or chain MNO is coiled.

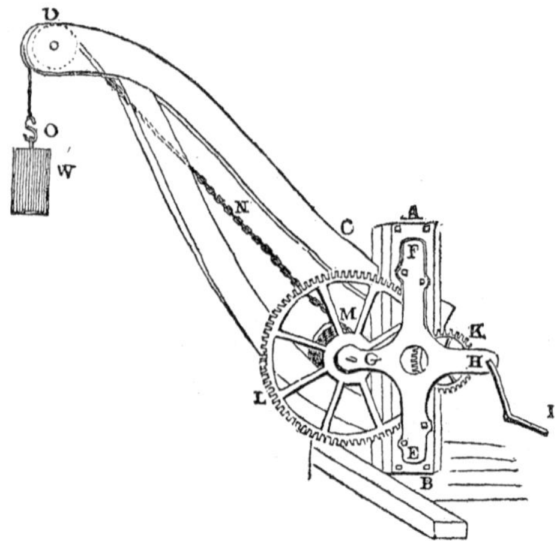


Ex. If the length of the winch be 18 in.; the diameter of the wheel K 10 in.; of the wheel L 24 in.; of the pinion attached to the winch 4 in.; of that upon the axle of the wheel K 4

---

White's pulley. Its advantages over the other forms. Diameters of the wheels.

**¶ 133.** Crane. The parts of which it is composed. Application of the power. How to increase the power.

in.; and of the barrel M 8 inches; what force will be exerted at W, by a power of 500 pounds applied at the winch ?

*Ans.* 33750 lbs.

## Hunter's Screw.

¶ **134.** If the power of the screw be increased, by diminishing the distance between the threads, the strength of the threads will be so diminished, that a slight resistance will tear them from the cylinder. This inconvenience is removed by *Hunter's Screw*, which consists of two screws upon the same cylinder, the threads being of unequal fineness. The threads may have any strength and magnitude, the efficacy of the screw depending not upon the size of the threads, but upon the difference between the distances of the threads of the two screws.

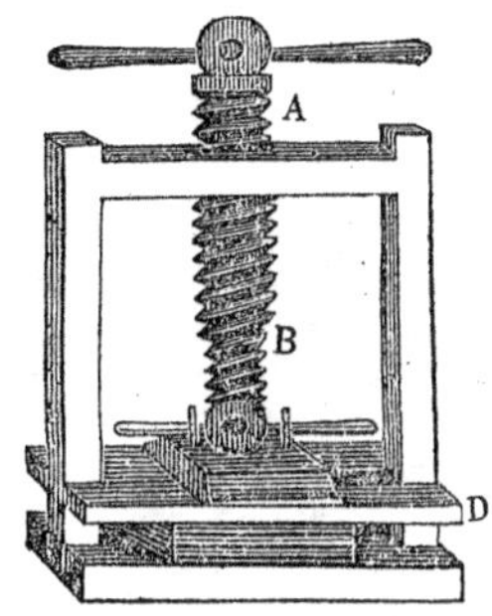


Ex. The screw A contains 15 threads to the inch, and the screw B 16; how far will the board D be depressed by one revolution of the screw ? If the lever which passes through the head of the screw A, be 21 inches long, and it be turned by a power of 50 pounds, what power will be exerted upon the board D, making a deduction of 52 per cent. from the power for friction ?

*Ans.* 380160 lbs.

## The Endless Screw.

¶ **135.** *The Endless Screw* consists of a screw combined with a wheel and axle in such a manner, that the threads of the screw work with the teeth of the wheel.

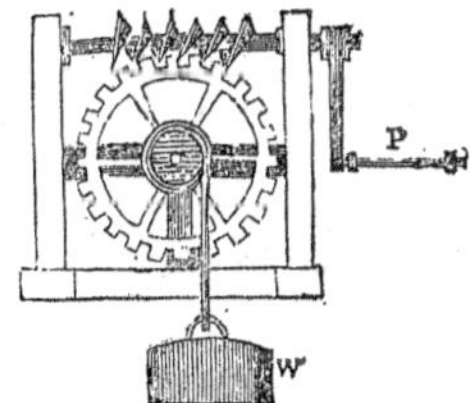


Ex. The winch is 14 inches long, the threads of the screw are $\frac{1}{2}$ inch apart, the radius of the wheel is 12 inches, and of

---

¶ **134.** Manner of increasing the power of the screw. The result of diminishing the distance between the threads. Hunter's screw.

¶ **135.** Endless screw.

the axle 4 inches; what is the ratio of the weight to the power? What power applied at P, will be sufficient to balance a weight of 436 pounds suspended at W? *Ans.* $1\frac{43}{66}$ lbs.

---

## Pumps.

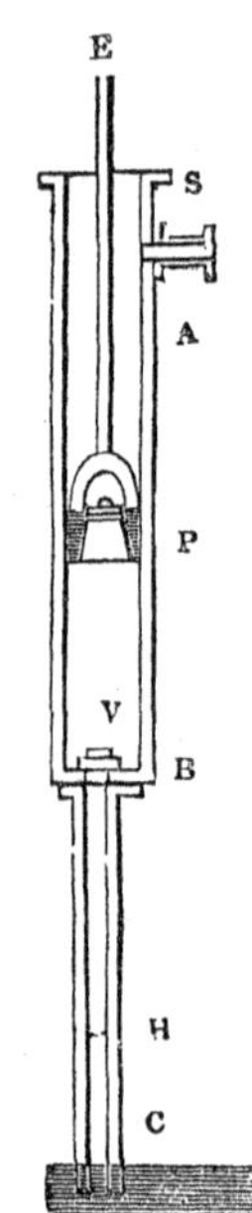


**¶ 136.** *A Pump* is a machine for raising water. The *Common* or *Suction Pump*, a section of which is represented in the diagram, is the one used for common household purposes. AC is a *pipe* of any convenient length, the lower end of which reaches below the surface of the water in the well or reservoir. The part AB of the pipe is commonly of greater diameter than the part CH. V is a *valve* opening upwards. P is a *piston* moved by the *rod* E. In this piston is also a valve opening upwards. To the upper end of the rod E, is attached the end of the short arm of a lever, the end of the long arm being the point at which the power is applied.

Ex. A cubic foot of water weighs $62\frac{1}{2}$ pounds. Suppose the piston P to be at the bottom of the pipe AB; the distance from B to A, 12 feet, to be filled with water; the diameter of the pipe AB to be 5 inches; the long arm of the lever or handle to be 30 inches; the short arm 5 inches; and the friction to be 10 per cent. What power applied at the end of the long arm of the handle, will be required to work the pump? If the power move through a vertical distance of 50 inches, what quantity of water will be discharged at one stroke of the lever?

*Ans.* { Power, 18 lbs. 11·9+ oz.
Water, $163\frac{5}{8}$ cu. in.

---

¶ **136.** Pump. Explanation of the common or suction pump.

## The Hydrostatic Press.

**¶ 137.** *The Hydrostatic Press* is a machine by which an enormous force of pressure is obtained through the medium of water. It consists of a short and very strong *pump barrel*, with a *solid piston*, C. To this is attached the *rod* D, mounted with the *crosspiece* E, which is pushed upwards against the thing to be compressed. AB is a small pump, called a *Forcing Pump*, which drives or forces water into the pump barrel C, and thus produces the pressure. If the small pump have only $\frac{1}{100}$ as great an area as the large barrel, a pressure of 1 pound in the pump AB, will produce a pressure of 100 pounds at E. But here, as in all other cases in machinery, what is gained in force, is lost in distance. See ¶ 118.

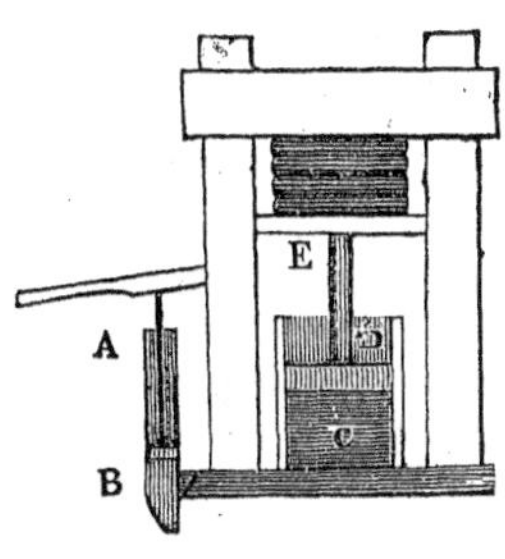


NOTE. The hydrostatic press is also called the *Hydraulic Press*, and sometimes, from the name of the inventor of its present form, *Bramah's Press*.

Ex. The forcing pump AB is 2 inches in diameter, and the barrel C 18 inches; the piston of the forcing pump is worked by a lever 25 inches long, the fulcrum being at one end; and the piston rod is attached to the lever at the distance of 5 inches from the fulcrum. What pressure will be produced at E, by a power of 140 pounds applied at F, no allowance being made for friction? *Ans.* 56700 lbs.

---

## Methods of Applying Power to Machinery.

**¶ 138.** It has been remarked, ¶ 125, that the power may be applied to a machine by gravity, animal strength, wind, water, steam, &c.

The application of gravity and of animal strength as moving powers, have already been made in the previous estimates upon machinery. But the two are not unfrequently combined, as will be seen in

---

¶ 137. Hydrostatic press. Explanation of the parts of which it consists. Illustration of its power. Note.

¶ 138. Topic. Reference to ¶ 125. Gravity and animal strength.

## The Tread-Mill.

**¶ 139.** *The Tread-Mill* consists of a wheel, usually about 5 feet in diameter, and 16 feet long. The circumference is furnished with 24 steps, on which prisoners are made to work. Several prisoners work together upon the same wheel, each treading in a separate compartment. It will readily be seen that the weight of several persons upon the same side of the wheel, will set it in motion. When the wheel is once in motion, the prisoner has no alternative, but must keep treading. He is assisted in a degree, however, by a *Hand-rail* before him, as shown in the diagram.

Ex. A wheel 5 feet in diameter, is worked by 5 men, whose average weight is 130 pounds. The wheel is connected with gearing, which moves a weight 4 feet while any point in the circumference of the wheel moves 1 foot. If the power be uniform, and the wheel revolve twice in a minute, what is the power of the machine, making no allowance for friction?

*Ans.* $\frac{4615}{7458}$ of 1 horse power.

**¶ 140.** The tread-mill is sometimes so constructed as to be moved by the weight of animals walking on an inclined plane, as shown in the diagram.

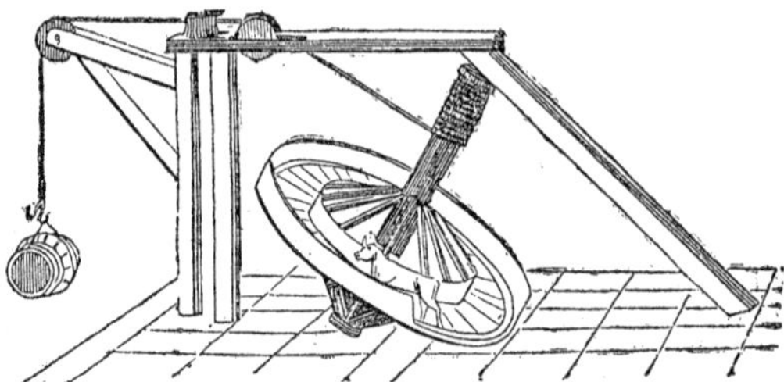

Ex. The diameter of the wheel represented in the diagram is 18 feet, and of the axle 16 inches. The rope, which coils round the axle, runs over the crane, and sustains the weight, is 2 inches in diameter. If an animal exert a force of 400 pounds

---

¶ **139.** Tread-mill. Its construction and use.

at the circumference of the tread-wheel, what weight may be raised at the end of the rope, deducting 25 per cent. from the power, for the friction of the machine? *Ans.* 3600 lbs.

NOTE. Wind has been employed to some extent as an agent or moving power to machinery. It has in most cases been superseded by animal, water, and steam powers, and is now seldom employed.

## Water Wheels.

¶ **141** *A Water Wheel* is a wheel turned by the force of running water. Water wheels are variously constructed, according to the circumstances under which they are intended to act. They are of three kinds, viz.:

1st. *The Overshot Wheel* is one in which the water is brought over the top of the wheel, received in buckets, and by its weight causes the wheel to revolve. It is employed chiefly where the stream affords but a small supply of water; but in no case can it be employed, unless the descent of the stream be somewhat rapid.

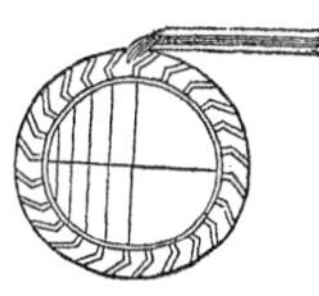

NOTE 1. The action of the water as a moving power is dependent upon the principles given in ¶ 130.

2d. *The Undershot Wheel* is one in which the water strikes the float boards below the axle. Its power depends upon the size of the wheel, and the size and velocity of the stream. It is employed chiefly where the supply of water is abundant, but the banks of the stream do not admit of a dam of sufficient hight to employ the overshot wheel to advantage.

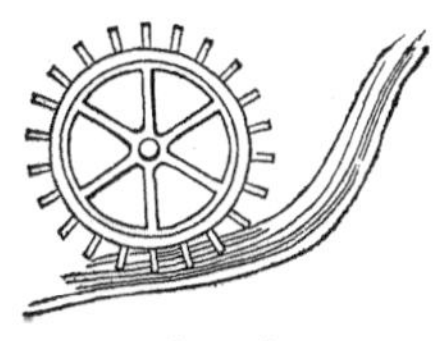

3d. *The Breast Wheel* is a combination of the overshot and the undershot wheel, since its force is obtained partly from the *weight*, and partly from the *velocity* of the water. It is employed chiefly where the supply of water is not sufficient to give the required force to an

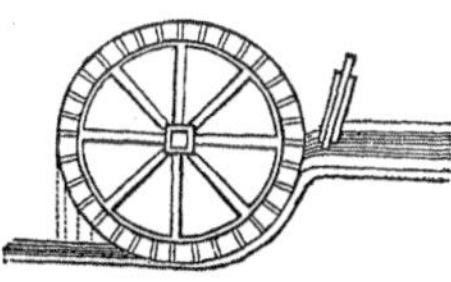

¶ **141.** Water wheel. The overshot wheel. Where employed. Note 1. The undershot wheel. Where employed. The breast wheel. Where employed. Note 2.

undershot wheel, and the banks of the stream do not admit of a dam of sufficient hight to employ an overshot wheel.

NOTE 2. With a given stream of water, under favorable circumstances, the power of the overshot wheel is about twice that of the breast wheel; and the power of the breast wheel from 2 to 5 times that of the undershot wheel.

NOTE 3. For an explanation of the methods of generating steam power, and applying it to machinery, the pupil is referred to Dr. Lardner's Lectures on Science, Literature, and Art; and to any of the various treatises on Natural Philosophy. And as a study in mechanics and machinery, the locomotive and stationary steam engines furnish the pupil with a greater variety of combinations and applications of the mechanical powers, than almost any other machines.

# OLMSTED'S RUDIMENTS OF NATURAL PHILOSOPHY AND ASTRONOMY. 1 vol. 18mo.

For common Schools and younger classes in Academies.

☞ Each part also bound by itself and sold separately.

This beautiful little volume is the latest of the series of text-books in Natural Philosophy and Astronomy prepared by Professor Olmsted. Its leading object is to afford an easy explanation of *those* truths of these sciences, which are most important to be known by mankind in general, being truths of the greatest practical utility. No similar work, it is believed, ever contained, in the same compass, a greater amount of useful and interesting matter. This is rendered easy and intelligible by familiar illustrations and expressive diagrams, and is adapted to the comprehension of young learners, to a degree which can be attained by those only who, like the author, have had great experience in teaching.

The approbation which this work has met with from teachers, is indicated by its having, within ten years of its first publication, passed through nearly twenty editions. On account of its simplicity of style and happy way of illustrating profound truths, it has been published in the form of *raised letters* for the use of the blind in the Massachusetts Asylum, at Boston, and has been widely circulated by the American Board among the Missionary Schools in distant parts of the earth.

The publishers take pleasure in submitting the following testimonials from qualified judges.

*From Cyrus Mason, Professor in New York City University, and Rector of the University Grammar School, and Lewis H. Hobby, Head Master.*

"We are not accustomed to give testimonials of our approbation of books used in the Grammar School; but we are constrained to make an exception in favor of the 'Rudiments of Natural Philosophy and Astronomy.' We have used this book from the day of its publication, with increasing pleasure to ourselves, and advantage to our pupils. It is pre-eminently adapted to the work of public instruction; clear, methodical, comprehensive and satisfactory; incapable of being used by a master who does not understand it, or of being recited by a pupil who has not comprehended its meaning.

"In the preparation of this book, Professor Olmsted has made himself a benefactor of the schools of our country; and we cannot but hope that its wide and early circulation will fully reward his labors."

---

*From the Journal of St. Paul's College, College Point, Long Island.*

"Of the publications of 1844, we have received the 'Rudiments of Natural Philosophy and Astronomy, designed for the younger classes in Academies and for Common Schools,' by Professor Olmsted, of Yale College, for which the author will accept our thanks. It is suited exactly to our purpose. For years past we have been obliged to depend upon oral instruction, for the lack of a good text book, and have, in common with many others, wished that some accomplished gentleman would undertake to furnish us a book calculated for beginners. It is generally supposed that elementary works can be prepared by persons of partial acquirements, or that they require less research than works of a higher order. In one sense this is true; but we maintain that a work adapted to the youthful mind, cannot be written by any other than by him who perfectly understands the subject in its higher departments. It is one of the signs of better times, that men of high talent are willing to give their attention to the preparation of elementary treatises.

"The work which has suggested these remarks, is judiciously divided into short paragraphs, and filled with neat and useful diagrams. We recommend it to the attention of all teachers in schools and academies."

---

*From the pen of a distinguished Clergyman (Albert Barnes.)*

"This is the title of a book (Rudiments, &c.) which has evidently been prepared with much care, and which is intended to be adapted to promote a very important object in schools and academies. Professor Olmsted has prepared on the same general subject, a Treatise on Natural Philosophy, in 8vo., a Treatise on Astronomy, in one vol. 8vo, a School Philosophy, and a School Astronomy, which have been received with great favor by the public, and which have passed through numerous editions. The little work, whose title is given above, completes his plan, by adapting this kind of instruction to primary schools. He has some rare qualifications for the composition of such a work. Besides his eminence in this department of instruction in Yale College, and his entire familiarity with his subject, he has had the advantage of having been himself a teacher in a common school and in an academy, and of thus becoming acquainted with the best method of approaching the youthful mind. The work is prepared with great care, with eminent ability and judgment, and is well adapted to interest the class of youth for whom it is intended.

"The writer of this notice, in commending this work to the favorable regards of the public, cannot but be struck with the difference between such a work and any one to which he had access in receiving his education, now almost forty years ago. At that time, no youth in the land had the advantage of a book so admirably adapted at once to cultivate the powers of reflection and investigation, and to interest the mind on the subjects of great importance to those entering on life. Nor does he now know of any work of this description, at once so comprehensive and so clear; so full of important principles of science, and so attractive to the youthful mind. Its introduction into the schools of this city, and the schools and academies of this commonwealth, he would regard as a circumstance arguing most favorably for the promotion of the best interests of education. Indeed, many a man who graduated at College, and who has entered on his professional life, would find it a work in which he would be greatly interested and profited." —*Philadelphia North American.*

"This we consider among the most valuable offerings of the series of the Professor to the youth of his country. Although professedly designed for the younger classes in academies and common schools, it is written with great ability and judgment, and is adapted to interest young persons of cultivated minds, and to secure their study."—*New York Evangelist.*

In addition to the foregoing recommendations, the publishers are in possession of others, also commendatory, from individuals in different parts. All concur in expressing high approbation of this work as eminently adapted to diffuse a taste for the studies of Natural Philosophy and Astronomy among the rising generation.

## M'CURDY'S FIRST LESSONS IN GEOMETRY:

Adapted, in connexion with the Charts of Geometry, to the use of Public Schools and Academies. By *D. M'Curdy* of Washington, D. C. 12mo. Price 25 cents.

## M'CURDY'S TWO CHARTS OF GEOMETRY:

On Rollers, size 34 by 48 inches; a sett, $5 00.

## M'CURDY'S EUCLID'S ELEMENTS, OR SECOND LESSONS IN GEOMETRY: 12mo. 37½ cts.

## COFFIN'S SOLAR AND LUNAR ECLIPSES

Familiarly illustrated and explained, with the method of calculating them, as taught in the New England Colleges. By *Jas. H. Coffin.* 8vo.

## COFFIN'S CONIC SECTIONS AND ANALYTICAL GEOMETRY. 8vo.

By *Jas. H. Coffin*, of Lafayette College, Pa.

The Demonstrations in this work are considered more simple and concise than in other treatises.

## ABERCROMBIE'S INTELLECTUAL PHILOSOPHY:

With additions, explanations and questions, to adapt the work to the use of schools and academies, by Jacob Abbott. Revised edition. 12mo.

---

## ABERCROMBIE'S MORAL PHILOSOPHY:

With additions and explanations, and also questions for the examination of classes, by Jacob Abbott. Revised and enlarged. 12mo.

---

## PRESTON'S TREATISE ON BOOK-KEEPING:

Comprising both Single and Double Entry; enlarged and improved edition. 8vo.

---

## PRESTON'S SINGLE ENTRY BOOK-KEEPING:

Adapted to the use of Retailers, Farmers, Mechanics and Common Schools. 8vo.

---

## PRESTON'S DISTRICT SCHOOL BOOK-KEEPING,

Printed on thick demy writing paper, for practice. An excellent work for beginners. Quarto. Price 25 cents.

---

## WHELPLEY'S COMPEND OF HISTORY.

Enlarged, with Questions, by *Joseph Emerson.* 12mo.

---

## HALE'S UNITED STATES PREMIUM HISTORY,

Revised and enlarged, with maps and questions, adapted to the use of Schools. Large 18mo.

---

## MURRAY'S SEQUEL TO THE ENGLISH READER.

12mo.

*

## ABBOTT'S MOUNT VERNON JUNIOR READER.

18mo.

---

## ABBOTT'S MOUNT VERNON MIDDLE CLASS READER. 18mo.

---

## ABBOTT'S MOUNT VERNON SENIOR READER.

12mo.

---

## ABBOTT'S MOUNT VERNON ARITHMETIC.

Part 1. Elementary. 12mo.

---

## ABBOTT'S MOUNT VERNON ARITHMETIC.

Part II. Fractions. 12mo.

*Mount Vernon Arithmetic, by Jacob Abbott. Part* I.—*Elementary. Part II—Fractions.*

Teachers are invited to examine these works, which are constructed on a plan materially different from that of the books in common use. The various principles involved in the several arithmetical processes are unfolded in a very clear and gradual manner, each being illustrated by a great number of examples, of nearly equal difficulty, so that the pupils have a full supply of materials for practice, without the necessity of continually resorting to the teacher for explanation and aid. Thus, the labor of the teacher is greatly abridged, abundance of pleasant and profitable employment is furnished for the pupils, and the knowledge which they acquire of the subject is of the most thorough and practical character.

These books contain also a series of exercises on an entirely new plan for teaching the art of adding up columns of figures with rapidity and correctness. These exercises are to be practised in classes, and are found very successful wherever they are introduced.

---

## ABBOTT'S COMMON SCHOOL DRAWING CARDS.

A new article; 40 Cards of Landscapes and Flowers, with Directions for Drawing on the back of each Card. In a neat case. Price 50 cents.

## ADDICK'S ELEMENTS OF THE FRENCH LANGUAGE:

An elementary practical work for learning to speak the French Language, expressly adapted to the capacity of Children; containing a table of the French sounds and articulation, with corresponding lessons in pronouncing and reading. 12mo.

---

## GIRARD'S ELEMENTS OF THE SPANISH LANGUAGE:

An elementary practical Work for learning to speak and write the Spanish Language, from the method of Dr. J. H. P. Seidenstuecker. By *J. F. Girard.* 12mo.

---

## KIRKHAM'S ENGLISH GRAMMAR:

In Familiar Lectures; containing a new systematic order of Parsing, exercises in False Syntax, and a system of Philosophical Grammar, in notes; to which are added an Appendix, and a Key to the Exercises. 12mo.

---

## THE GOVERNMENTAL INSTRUCTOR,

Or a Brief and Comprehensive view of the Government of the United States, and of the State Governments, in easy lessons, designed for the use of schools, by *J. B. Shurtleff.* Revised edition. 12mo. Price 37 1-2 cents.

---

## DAY'S MATHEMATICS. 8vo.

---

## OUR COUSINS IN OHIO:

By *Mary Howitt.* From the Diary of an American mother. Large 18mo.

## *Adams' Arithmetical Series.*

### I.—PRIMARY ARITHMETIC, OR MENTAL OPERATIONS IN NUMBERS: 18mo.

Being an Introduction to Adams' New Arithmetic Revised edition.

### II.—ADAMS' NEW ARITHMETIC, REVISED EDITION: 12mo.

Being a revision of Adams' New Arithmetic; first published in 1827.

### III.—KEY TO THE REVISED EDITION OF ADAMS' NEW ARITHMETIC. 12mo.

### IV.—MENSURATION, MECHANICAL POWERS, AND MACHINERY:

The principles of Mensuration analytically explained, and practically applied to the measurement of lines, superfices and solids; also, a philosophical explanation of the simple mechanical powers, and their application to machinery. Designed to follow Adams' New Arithmetic. Preparing.

---

## JUVENILE DRAWING BOOK.

John Rubens Smith's Juvenile Drawing Book, being the rudiments of the art, explained in a series of easy progressive lessons, embracing drawing and shading; comprised in numerous copperplate engravings, with copious letter-press instructions. Large quarto.

---

## THE OXFORD DRAWING BOOK:

Containing Progressive Lessons on Sketching, Drawing, and Colouring, Landscape Scenery, Animals, and the Human Figure. Embellished with 100 lithographic drawings. Quarto.

## DYMOND'S ESSAYS,

## ESSAYS ON THE PRINCIPLES OF MORALITY,

And on the Private and Political Rights and Obligations of Mankind. By *Jonathan Dymond.* Complete, stereotype edition, 12mo., 576 pages. Price, 50 cents.

"I have long entertained a very favorable opinion of Dymond's Essays. The rules that he prescribes are strict, but so also are those of the Universal Governor. His spirit is kind and catholic, and I think that no one can attentively read his book without deriving profit from it. You have rendered a service to the moral interests of the country by giving this edition of so valuable a work to the public." JOHN PIERPONT.

"The work is one which should find a place in every School District Library, as an invaluable compend of pure truth and sound ethics."

SAMUEL S. RANDALL.

*Secretary's Office, Department of Common Schools.*

"I think its circulation must have a very salutary tendency."

JACOB ABBOTT.

"This very excellent work is not so well known, or so extensively read, as it should be. It breathes the purest Christian spirit, and the most high-toned morality. It deserves a place as a household book, on every one's table."—*N. A. S. Standard.*

"This volume, beyond all comparison, is the best work upon the themes generally called 'Moral Philosophy' in our language—and the person who has not studied it, is not a master of that most noble science. This is a neat edition, and is more valuable and useful on account of the Index which has been added. We are rejoiced to see that the work is stereotyped, and therefore that its *cheapness* will ensure its dissemination."—*Christian Intelligencer.*

NO CROSS NO CROWN. By *William Penn.* 12mo.

## THE IMITATION OF CHRIST:

In three books. By *Thomas à Kempis.* Translated from the Latin, by *John Paine*, with an Introductory Essay by *Thomas Chalmers.* Stereotype edition; 12mo.

## MEMOIRS OF SAMUEL FOTHERGILL.

With Selections from his Correspondence, &c. By *Geo. Crossfield.* 8vo. Price, in muslin $2 00, sheep $2 25.

---

## THE NEW TESTAMENT.

Small 8vo., large type; price 50 cents.

---

## JOURNAL

Of the Life, Gospel Labors, and Christian Experiences of *John Woolman,* to which are added his last Epistle, and other Writings. 12mo. Price, 37 1-2 and 50 cents.

---

## JOURNAL OF THE LIFE AND GOSPEL LABORS OF DAVID SANDS:

With extracts from his correspondence. 12mo. Price 62 1-2 cents.

---

☞ *A liberal discount made to those who purchase in quantities.*

---

www.ingramcontent.com/pod-product-compliance
Lightning Source LLC
LaVergne TN
LVHW021413110826
845150LV00007B/1909

*9781425510428*